W9-AGP-341

Power System Operation

Energy control center of a large power system (Pacific Gas and Electric Company).
NOTE: *Control console center left with EHV dynamic board,*
and computer equipment in background
beyond glass partition.

Power System Operation

Robert H. Miller
Consulting Engineer

SECOND EDITION

McGraw-Hill Book Company

New York St. Louis San Francisco Auckland Bogotá
Hamburg Johannesburg London Madrid Mexico
Montreal New Delhi Panama Paris São Paulo
Singapore Sydney Tokyo Toronto

Library of Congress Cataloging in Publication Data

Miller, Robert H. (Robert Herschel), date.
  Power system operation.

  Bibliography: p.
  Includes index.
  1. Electric power systems—Management.
  2. Electric power systems—Load dispatching.   I. Western
  Systems Coordinating Council.   II. Title
  TK1005.M498   1983   621.319   82-24959
  ISBN 0-07-041975-2

Copyright © 1983 by McGraw-Hill, Inc. All rights reserved.
Printed in the United States of America. Except as permitted
under the United States Copyright Act of 1976, no part of this
publication may be reproduced or distributed in any form or by
any means, or stored in a data base or retrieval system, without
the prior written permission of the publisher.

1234567890   KGP/KGP   89876543

ISBN 0-07-041975-2

*The editors for this book were Diane Heiberg, Alice V. Manning,
and Peggy Lamb, the designer was Naomi Auerbach, and the
production supervisor was Sally Fliess.
It was set in Baskerville by J. M. Post Graphics, Corp.
Printed and bound by The Kingsport Press.*

To Power System Dispatchers

Contents

Preface

This book was originally produced in 1970, under the auspices of the Western Systems Coordinating Council, primarily to provide power system operators with material useful in power system operation, including information on interconnection. Apparently it did fill a need, at least to some extent, as it has received much greater acceptance than was originally anticipated, both domestically and in foreign countries.

Because of the many advances in control technology and system protection, and the changes in fuel costs for power generation, some sections of the first edition were in need of revision and expansion to bring the text up to date. It is for these reasons that this new edition has been prepared.

There has been no change in intent, however. The objective is still to provide, in as simple terms as possible, a description of the functioning of power systems and their control. As in the first edition, the use of advanced mathematics has been avoided, and appendix sections have included the development of the required mathematics in sufficient detail to meet the requirements of the readers.

It is sincerely hoped that the material presented in this book will be of value to power system operators, and that it will enable them to better understand the behavior of power systems under both normal and abnormal conditions.

The revised text has been reviewed by Robert F. Wolff of *Electrical World* and Sheldon Strauss of *Power Magazine*. Thanks are also due to Peggy Lamb of McGraw-Hill for her suggestions on construction and arrangement, and to Diane Heiberg, who has had the responsibility of editing the book.

R. H. MILLER

Preface to the First Edition

This book has been prepared under the auspices of the Western Systems Coordinating Council in an effort to provide material that would be useful to power system dispatchers in better understanding the principles of electric power system operation. To the extent possible the use of mathematics has been avoided. Where mathematics has been required, appendix sections have been included to develop the basis for the mathematical applications.

A considerable portion of the manual has been devoted to power system control, economics, and interconnected operation. Although these topics are well understood by engineers, there is little information conveniently available for power system operators not having formal engineering training.

Procedural information for clearing lines and equipment has been purposely avoided since practices on individual power systems vary and it is believed that such procedures can best be developed by the systems or by mutual agreement for interconnected operation.

The manual has been reviewed by various representatives of the Western Systems Coordinating Council and their helpful suggestions are appreciated. Particular thanks are due Mr. William Bosshart of the Bonneville Power Administration, Mr. Clyde Reikofski of the United States Bureau of Reclamation, and Mr. K. K. Dols of the Northern States Power Company for their detailed reviews of the material in the manual and their suggestions and constructive criticisms of the text material, and Mr. Harry McMasters of the Pacific Gas and Electric Company for his careful reading and editorial suggestions. Thanks are also due Mrs. Lorrette Hopson and Mrs. Esperanza Martinez of the Pacific Gas and Electric Company for their patience in typing the drafts of the manuscript.

R. H. MILLER

Introduction

Although the basic principles of power system operation do not change, there have been significant developments in the tools available to power system operators since the first edition of this book was issued in 1970. Consequently, it seems appropriate to provide at least some discussion of these tools and their applications.

The basic purpose of the book has not changed—that is, to discuss in relatively simple terms how power systems operate and some of the problems that power system operators meet in their day-to-day operations. In general, the book attempts to consider system behavior and the effects of operations on a system rather than the functioning of specific types of power system equipment.

Because of the wide acceptance and application of supervisory control and data acquisition (SCADA) systems, a new chapter has been added describing such systems and some of their applications.

The section on power system protection has also been somewhat expanded to include directional-comparison and transfer trip relaying systems.

Other minor changes in the text and its arrangement have been made, in the hope that it will be more logical and an improvement over the original edition.

The discussions of the various items in the book are necessarily brief, as the intention has been to produce not a detailed textbook, but rather a ready source of at least some information on topics of interest to power system operators.

Basic Principles

1

Before the factors affecting the behavior of power systems are discussed, some basic electric circuit theory will be presented, since the behavior of all electric circuits and machinery is affected by the electrical components making up the circuits. A knowledge of these fundamental concepts is essential before power system behavior can be understood. All electric circuits contain resistance, inductance, and capacitance. The combination and proportion of these elements in a circuit determine the behavior of the circuit.

RESISTANCE

Resistance can be defined as the element of an electric circuit that limits the flow of current in the electric circuit. Electric energy flowing in a circuit containing resistance is converted to heat energy proportional to the square of the current. The energy consumed in a circuit containing resistance is equal to (amperes)^2 × resistance (ohms) watts, or power consumed = I^2R watts. The behavior of a circuit containing only resis-

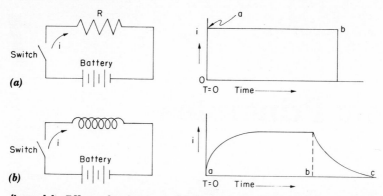

figure 1-1 Effects of resistance and inductance on change of flow of current. (*a*) In a circuit containing only resistance, when the switch is closed at $T = 0$, current rises instantaneously to its final value i at point a, and when the switch is opened at time b, it decreases immediately to zero. (*b*) In a circuit containing inductance, when the switch is closed at a ($T = 0$), current rises over a period of time until it reaches a final value i determined by the resistance and inductance in the circuit. When the switch is opened at time b, the current decays to zero over a period of time, again determined by the resistance and inductance in the circuit.

tance is shown in Fig. 1-1*a*. The amount of resistance in a circuit is affected by the resistivity of the material conducting a current. It varies inversely with the cross-sectional area of the material and directly with its length. Most metals are relatively good conductors of electricity; that is, they have a low resistivity. Other materials, such as wood, glass, or rubber, have a very high resistivity and are classified as insulators. Liquids and gases may have high or low resistivity, depending on temperature and various other factors.

INDUCTANCE

Inductance can be defined as the element of an electric circuit which opposes sudden increases or decreases of current flowing in the circuit. Inductance stores energy in a magnetic field when current is increased and delivers energy back to the circuit when current is decreased. The amount of energy returned to a circuit would exactly equal the amount of energy stored if there were no resistance losses.

The effect of inductance in a circuit is to delay current changes. Figure 1-1 shows the effect of resistance and inductance in a circuit.

After the initial closing of the switch, the current increases exponentially to a maximum, limited by the resistance in the circuit, and will maintain this value until the switch is opened, when it will begin to decrease exponentially. In an ac circuit the voltage is continuously changing. Consequently, the current will also change, but at a time somewhat delayed, because, as noted above, an inductance opposes a sudden change in current. As a result, the current in a circuit containing inductance is said to "lag" the voltage. One complete cycle is 360 electrical degrees, and in a circuit with pure inductance the current will lag the voltage by 90 degrees. Practical circuits always contain some resistance, so that in such a circuit the current will lag the voltage, but at an angle of less than 90 degrees, depending on the ratio of resistance to inductive reactance.

Coils of a conductor are primarily inductive but always contain some resistance. The amount of inductance is proportional to the number of turns and to the material used in the core. Magnetic materials such as iron greatly increase the inductance of a coil as compared with an air-core coil. Straight conductors, such as transmission lines, also have inductance proportional to the length of the line and the size and spacing of the conductors. This factor is very important in determining the behavior of long transmission lines.

CAPACITANCE

Capacitance is the property of electric circuits through which electric energy is stored in an insulating medium (dielectric). Electrical capacitance exists when two conductors are separated by a dielectric, which may be air, paper, mica, porcelain, glass, or other insulating material. The effect of capacitance in a circuit is shown in Fig. 1-2.

As has previously been noted, all electric circuits contain electric resistance, inductance, and capacitance. In ac circuits, the effect of in-

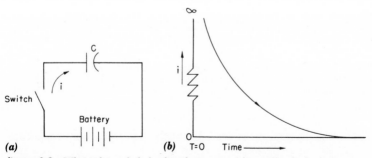

figure 1-2 When the switch is closed, current i immediately increases to a value limited only by the resistance in the circuit and then decreases to zero over a period of time as indicated in (*b*). If the switch is opened after i reaches zero, no change will occur, and an electric charge (electrostatic energy) will be left on the plates of the capacitor. With a perfect dielectric (i.e., one with no loss), the charge would be retained indefinitely. In an ac circuit, since voltage is continuously changing, the effect is alternately to charge the capacitor in one polarity, discharge it, and then charge it in the opposite polarity.

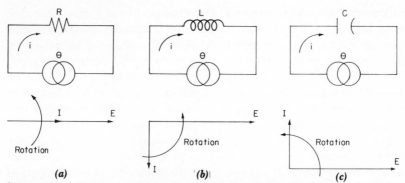

figure 1-3(a) Circuit containing only resistance; current and voltage are in phase. (b) Circuit containing only inductance; current lags voltage by 90°. (c) Circuit containing only capacitance; current leads voltage by 90°.

ductance is to delay changes in the flow of current, so that current is said to lag the voltage when the circuit is primarily inductive. In circuits that contain capacitance, the current required to charge the dielectric is greatest when the rate of voltage change is greatest and reaches zero when the rate of voltage change is zero. The result is that current in a primarily capacitive circuit leads the voltage. In a purely resistive circuit, current varies exactly in proportion to the voltage, and hence current and voltage are in phase. These conditions are shown in vector diagrams of Fig. 1-3.

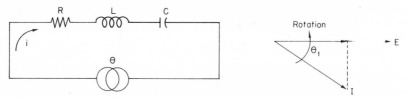

figure 1-4(a) Inductive reactance X_L exceeds capacitive reactance X_C, and current lags voltage by angle θ_1. The in-phase component (projection) equals $I \times$ cosine θ_1. Cosine θ is the power factor.

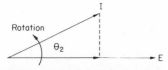

figure 1-4(b) Capacitive reactance X_C exceeds inductive reactance X_L. Hence current leads voltage by angle θ_2. The in-phase component equals I cosine $\theta = I \times$ power factor.

figure 1-4(c) Inductive reactance equals capacitive reactance, and the two components cancel. Current is in phase with the voltage and limited by resistance in the circuit. Power factor $= 1$.

The circuits shown in Fig. 1-3 are idealized, and, as previously noted, all circuits actually contain resistance R, inductance L, and capacitance C. In reality a circuit will always contain resistance, and if the inductive component L exceeds the capacitive component C, the current will lag the voltage, but at an angle of less than 90°. Conversely, if the circuit is primarily capacitive, the current will lead the voltage, but at an angle of less than 90°. In other words, inductance and capacitance have opposite effects in a circuit, and at a specific frequency the effects of inductance and capacitance will cancel. This is the condition known as resonance. In tuning a radio receiver, the coils and capacitors in the receiver are made resonant to the frequency of the station being received. At resonance, the inductive and capacitive reactances are equal, and current and voltage in the circuit are in phase (power factor = 1). These relations are shown graphically in Fig. 1-4.

REACTANCE

The terms "inductive reactance" and "capacitive reactance" have been introduced in the previous discussion. It may, at this time, be appropriate to mention the difference in the behavior of these two quantities. Both inductive and capacitive reactances are measured in ohms, the same unit used to measure resistance.

As has been previously pointed out, inductance tends to delay the change in the flow of current. It follows that if frequency is increased, the effect of inductance increases, and inductive reactance varies directly with frequency. Inductive reactance X_L is equal to 2π times the frequency times the inductance L, which is measured in henrys, or, mathematically, $X_L = 2\pi f L$ Ω. For 60 Hz, $X_L = 377L$ Ω.

It has also been pointed out previously that the effect of capacitive current is dependent on the rate of change of voltage. Consequently, capacitive current increases as frequency increases, or, in other words, the capacitive reactance X_C decreases as the current increases. Capacitance is measured in farads. Mathematically, capacitive reactance is expressed as

$$X_C = \frac{1}{2\pi f C} \quad \Omega$$

For 60 Hz,

$$X_C = \frac{1}{377C} \quad \Omega$$

The unity-power-factor case mentioned earlier occurs when $X_L = X_C$.

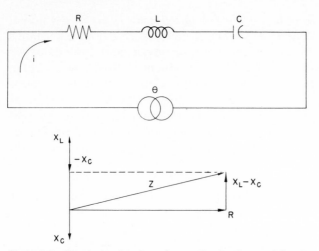

figure 1-5 Sketch showing impedance of a circuit containing R, X_L, and X_C. The total impedance is always the square root of the resistance squared plus the square of the difference of the inductive reactance and the capacitive reactance, or $Z = \sqrt{R^2 + (X_L - X_C)^2}$ ohms. Electrical engineers use a simpler notation, which unfortunately is called complex; it would express Z as $R + {}_j(X_L - X_C)$ where j is equal to the square root of -1 ($\sqrt{-1}$).

IMPEDANCE

Since all circuits contain resistance R, inductive reactance X_L, and capacitive reactance X_C, the effects of all these quantities on the flow of current in the circuit are seen. The total effect of R, X_L, and X_C is called impedance (Z) and is also measured in ohms. The impedance Z is the vector sum of the resistance and reactances in a circuit. This is illustrated in Fig. 1-5.

The preceding discussion has dealt with single-phase systems. The usual utility transmission system is three-phase, which complicates matters somewhat.

THREE-PHASE SYSTEMS

Without going through the proof, it can be stated that the power in a balanced three-phase circuit is the square root of 3 times the line voltage times the line current times the power factor, or power P $= \sqrt{3}\,EI$ power factor (watts).[1] In a star- or wye-connected system it can also be shown

---

[1]This assumes a balanced system, that is, a system in which the currents and voltages are equal in all the phases. Unbalanced phase currents and voltages present a more complicated problem. Power systems are normally balanced, so that the above statement is correct in these cases. Unbalanced systems are beyond the scope of this manual, and consequently will not be treated. Further treatment of unbalanced systems can be found in electrical engineering texts.

that line-to-line voltage is equal to $\sqrt{3}$ times line-to-ground voltage E, or

$$E_L = \sqrt{3}\,E \qquad \text{V}$$

and in a delta system line current is equal to $\sqrt{3}$ times phase current I, or

$$I_L = \sqrt{3} \times I \qquad \text{A}$$

Obviously the product of current and voltage, when current is not in phase with the voltage, is greater than these quantities multiplied by a power factor which cannot exceed 1.

REAL AND REACTIVE POWER

The apparent power (product of current and voltage) is equal to the square root of the sum of the squares of voltage times current in phase and voltage times current out of phase, or

$$E \times I = \sqrt{(E \times I \text{ in phase})^2 + (E \times I \text{ out of phase})^2} \qquad \text{VA}$$

Due to the large quantities handled in utility transmission and distribution systems, the terms kVA = voltamperes/1000 and MVA = voltamperes/1,000,000 are usually used. A simpler method of expressing the voltampere product is

$$EI = \sqrt{(\text{watts})^2 + (\text{vars})^2} \qquad \text{VA}$$

Diagrammatically the relationships described are shown in Fig. 1-6.

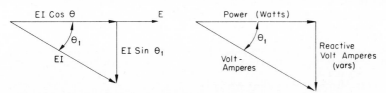

figure 1-6(a) Relationships of real and reactive components in a single-phase lagging-power-factor circuit (predominantly inductive).

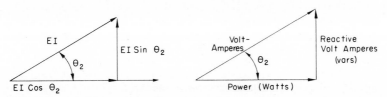

figure 1-6(b) Relationships of real and reactive components in a single-phase leading-power-factor circuit (predominantly capacitive).

The reactive component, that is, the component 90° out of phase with the power component, is referred to as voltamperes reactive, or var, or

$$\frac{\text{var}}{1000} = \text{kvar} \qquad \text{or} \qquad \frac{\text{var}}{1,000,000} = \text{Mvar}$$

The power factor of a circuit can readily be determined by dividing the power by the voltamperes.

$$\text{Power factor (PF)} = \frac{\text{power}}{\text{voltamperes}} = \cos \theta†$$

By referring to the diagrams of Fig. 1-6, it can readily be seen that whenever there is a reactive component, the line current is greater than it would be if the power factor were equal to 1. In other words, supplying the reactive (var) component in an ac electrical system causes an increase in current, which results in increased losses.

PROBLEMS

1. In the circuit shown in Fig. 1-1a, assume that the voltage of the battery is 10 V and the resistance R is 5 Ω. At time a, immediately after the switch is closed, the current in the circuit will be
     (a) Zero
     (b) 2 A
     (c) 5 A
     (d) 1 A
2. Immediately after the switch (Fig. 1-1a) is opened, the current will
     (a) Start to decay
     (b) Be zero
     (c) Be limited by the resistance R
3. In Fig. 1-1b, if the inductance is 2 H, the resistance 2 Ω, and the battery voltage 10 V when the switch is closed and steady-state conditions established, the current in the circuit will be
     (a) $\frac{1}{2}$A
     (b) Limited by the inductive reactance in the circuit
     (c) 5 A
4. If the condenser in Fig. 1-2a has a capacitance of 10 μF and a resistance of 0.01 Ω, and the voltage of the battery is 10 V, immediately after the switch is closed the current in the circuit will be
     (a) Limited by the capacitive reactance
     (b) 1000 A
     (c) 100 A
     (d) 9.9 A
5. After the condenser in Fig. 1-2a is fully charged, the current in the circuit will be
     (a) Limited by the resistance in the circuit
     (b) 1 A
     (c) Zero

†See Appendix 1 for development of trigonometric functions.

6. If a coil with an inductive reactance of 100 Ω at 1000 Hz is connected in series with a capacitor, in order for the circuit to be resonant at 1000 Hz, the capacitor should have a capacitive reactance of
    (a) 1000 Ω
    (b) 100 Ω
    (c) 1 Ω

7. If the circuit shown in Fig. 1-5 contains a resistance of 3 Ω, an inductive reactance of 5 Ω, and a capacitive reactance of 1 Ω, the total impedance of the circuit will be
    (a) 9 Ω
    (b) 3 Ω
    (c) 5 Ω
    (d) Capacitive

8. Indicating meters in a 60-kV transmission circuit show 4000 kW and 3000 kvar. The total kVA load of the circuit is
    (a) 5000
    (b) 7000
    (c) 1000
    (d) 4000

9. If the reactive voltamperes of a circuit are lagging, the power factor of the circuit can be improved by
    (a) Raising the voltage
    (b) Adding inductive reactance to the circuit
    (c) Adding capacitive reactance to the circuit

10. The power factor of a circuit is
    (a) The sine of the angle between the current and the voltage in the circuit
    (b) Always greater than unity
    (c) The ratio of the capacitive and inductive reactances in the circuit
    (d) The ratio of the power and the voltamperes in the circuit

2

Transfer of Energy in Power Systems

TRANSMISSION OF POWER

In electric power systems the power-generation equipment is usually located at some distance from the points of consumption of the power. Consequently it is necessary to transmit the energy from the locations where the power is produced to the points where it is used.

Electric power is transferred from generation to load by transmission and distribution lines. There is actually no difference between a transmission line and a distribution line except for the voltage level and power-handling capability. Lines referred to as transmission lines usually are capable of transmitting large quantities of electric energy over relatively great distances and operate at high voltages, of the order of 60,000 to 500,000 V or more. Distribution lines carry limited amounts of power over shorter distances and usually operate at voltages of the order of 2000 to 40,000 V.

In any case, the problem is to transfer electric energy economically and reliably from one location to another. This section will discuss some of the factors involved in the transfer of energy on electric power systems.

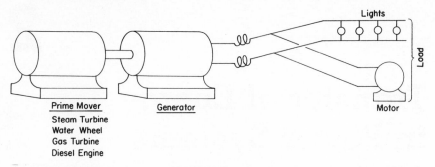

figure 2-1   Simple electric power system with single prime mover, generator, and load.

PRODUCTION OF ELECTRIC ENERGY

In the simplest case, only one prime mover, one generator, and a load are involved (Fig. 2-1).

The prime mover drives the generator, which produces electric energy. When a load, such as lights or a motor, is connected to the generator, the electric energy flows from the generator to the load. With the generator driven at constant speed and with no change in the generator field excitation, an addition to the electrical load will cause a drop in the speed of rotation of the prime-mover–generator combination and a reduction in the voltage supplied to the load. In order to bring the speed and voltage back to normal, more energy input to the prime mover is required. This is analogous to opening the throttle of a car to maintain speed when climbing a hill. In any case, the electrical output of the generator (watts, kilowatts, or megawatts) is determined by the connected load; this is true in both ac and dc systems. When two or more generators are connected together, as in a power system, the division of load between the generators is affected by several factors, which will be discussed in the following paragraphs.

DIVISION OF LOAD BETWEEN GENERATORS

In the dc case, control of the field voltage of the generators will serve to transfer load from one generator to another. Of course, the prime-mover inputs must be adjusted to provide the mechanical power needed to drive the generators at new load conditions. Figure 2-2 shows graphically the change in division of load between dc generators by field control.

The ac case is somewhat more complex. Power can be shifted between generators only by adjusting the inputs of the prime movers. Excitation

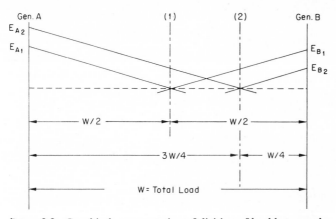

figure 2-2 Graphical representation of division of load between dc generators. (1) With a constant load W and with both generators adjusted to equal voltage $E_{A1} = E_{B1}$, assuming equal voltage drops, the loads are equally divided, with $W/2$ on each generator. (2) With generator A voltage increased to E_{A2} and generator B voltage reduced to E_{B2}, three-fourths of the load is on generator A and one-fourth on generator B.

(field-current) changes will cause reactive current flows between the machines but will not materially affect the real power division. Field-current adjustments with constant power input to the prime mover will affect the power factor of the generators but will not change the real power output. Figure 2-3 shows vector diagrams of a single-phase ac generator under several conditions of excitation, neglecting all impedance drops.

To summarize, the power output of a generator can be changed only by altering the power input to the prime mover. In the case of ac generators, field-current control alters voltage and current and their phase positions, changing the apparent power (product of current and voltage) but with no effect on the real power output. With dc generators, changing the field voltage will transfer load from one generator to another.

It should be stressed that the vector diagrams in Fig. 2-3 are greatly simplified by neglecting impedance drops and serve only to show the effect of field control on the phase position of voltage and currents.

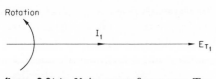

figure 2-3(a) Unity-power-factor case. Terminal voltage E_{T1} and current I are in phase. Power output = $E_{T1} \times I_1$ watts.

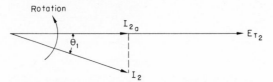

figure 2-3(b) Lagging-power-factor (overexcitation) case. With the generator field current increased, the terminal voltage E_{T2} is increased over the unity-power-factor case, and the current I_2 is also increased but out of phase lagging. The product of the projection of I_2 on the voltage vector E_{T2} is I_{2a}. This value of current multiplied by the machine terminal voltage is the real power output and is identical with the unity-power-factor case. The product of I_2 and E_{T2} is the apparent power (voltampere) output and is greater than in the unity-power-factor case. Real power is always determined by multiplying armature or line current times terminal voltage times the cosine (power factor) of the angle between the current and the voltage.

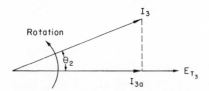

figure 2-3(c) Leading-power-factor (underexcitation) case. With generator field current reduced, the terminal voltage E_{T3} is reduced and current I_3 is increased but out of phase leading. Again the product of terminal voltage and the in-phase component of current is the same as in the unity-power-factor case.

ROTOR ANGLES OF AC MACHINES

As has been stated previously, the power output of a generator is changed only by altering the power input to the prime mover. The effect of increasing mechanical power input to a synchronous ac generator is to advance the rotor to a new position relative to the revolving electric field of the stator. (The concept of revolving fields is discussed in Appendix 3.) Conversely, a reduction of mechanical input will retard the rotor. In an idealized case of a machine with no losses and at no load, the field pole of the rotor would pass under the armature coils with no angular displacement and with no power output. With mechanical power input, the rotor will advance with respect to the stator, and electrical output

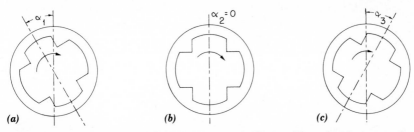

figure 2-4 (*a*) Rotor angle lagging armature winding position, electric energy absorbed, and mechanical energy produced (motor action). (*b*) Rotor angle zero, no electrical or mechanical input or output. (*c*) Rotor angle leading, mechanical energy absorbed, and electric energy produced (generator action).

will result. When a mechanical load is applied to the shaft of a machine, the rotor will retard with respect to the stator, and electric power will be absorbed, with mechanical power output. In other words, the angle of the rotor with respect to the revolving stator field determines whether a synchronous machine operates as a generator or as a motor. Figure 2-4 illustrates these conditions.

The greater the mechanical input to a generator, the greater the rotor angle leading and the greater the electrical output. The greater the mechanical load on a motor, the greater the rotor angle lagging and the greater the electrical input.

PARALLEL OPERATION OF SYNCHRONOUS AC GENERATORS

When synchronous ac generators are operated in parallel, it is necessary to increase or decrease the mechanical input to the prime mover of the machine whose load is being changed in order to increase or decrease load on one of the machines.

As a simple case, assume two identical generators supplying a constant load which is equally divided between the two generators. If it is desired to have machine *A* carry three-fourths of the load and machine *B* carry one-fourth, the mechanical input to machine *A* will be increased. At this time the machine will momentarily speed up so that its rotor angle is advanced by an amount sufficient to carry the new load. Simultaneously, the mechanical input to machine *B* will be reduced, and the machine will momentarily slow down until its rotor angle is retarded to a new position corresponding to the desired load condition.

Since the machines are synchronous, they will continue to run at the same average electrical speed with only the momentary acceleration and deceleration needed to stabilize at the desired load division. A complete

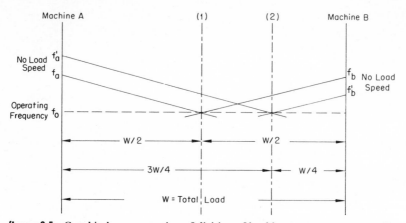

figure 2-5 Graphical representation of division of load between ac generators. (1) With a constant load W with the governors of both machines adjusted for speeds f_a and f_b at no load and separated. When paralleled and carrying load W, the governor speed droops will cause the speeds to decline to the operating frequency f_o. At this point the two speed curves intersect, with the machines sharing the load equally. (2) With the governor for machine A adjusted so that its no-load speed corresponds to f'_a, the generator will carry $3/4W$ at operating frequency f_o. Likewise, with the governor for machine B adjusted so that its no-load speed would correspond to f'_b, the generator will carry $1/4W$ at operating frequency f_o. Frequency f is proportional to speed of rotation of the machines.

description of what occurs electrically in the machines would require a vector diagram showing internal impedance drops, synchronizing current, and machine and terminal voltages, and is beyond the scope of this discussion. Suffice it to say that the machine with the greater leading rotor angle will carry the greater portion of the electrical load.

In practice, load division between machines is dependent on the speed characteristics of the prime-mover governors. Figure 2-5 illustrates the changing of load division between two alternators.

STABILITY

As discussed previously, if the mechanical input to the prime mover of a synchronous generator is increased, the rotor angle (torque angle) increases to a new position ahead of its former position. This is true up to a point at which the torque from the mechanical input exceeds the magnetic forces holding the machine in synchronism. When this condition is reached, the machine will be pulled out of step with the other machines with which it is paralleled. Figure 2-6 shows this relationship.

The concept of power angles in ac machines has been discussed at considerable length for two reasons: It is important in understanding the behavior of machines in a power system in which many machines are interconnected, and in understanding power flow on transmission lines and between interconnected systems. The important concept to remember is that the greater the advance of the power angle, the greater the flow of power, up to the maximum at which instability occurs.

When an electrical load is connected to a power source (one or more generators) by means of a transmission line, energy flows from the generators to the load. In an ac system, reactance is always present in transmission lines. The line behaves as though it is made up of series inductance and shunt capacitance distributed throughout its length with a relatively small series resistance component.

In order to transfer power from a generating source to a load over a transmission system, it is necessary for the load current to flow through the reactance resulting from the series inductance of the line. As a result, there is always a phase shift between the sending and receiving ends of the line, with the phase shift increasing as line loading is increased. As in the case of loading ac generators, power transfer increases up to a maximum determined by the line length, conductor size, and spacing, all of which affect the reactance of the line. When the maximum power transfer load is reached, any further increase will cause the angle to increase into the unstable region; an out-of-step condition will follow.

The variation of angular displacement with load is shown in Fig. 2-7.

Stability limits of a line can be increased, within limits, by the addition of series capacitors which compensate for the series inductive reactance of the line. This is being done on the Pacific Intertie.

From Fig. 2-7, it can be seen that if a line is loaded near the stable limit and additional power is required to flow on it, for example, following the loss of another line or due to a fault, the angular displacement

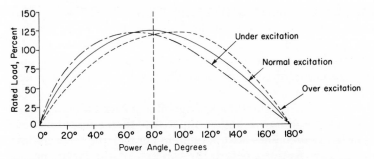

figure 2-6 Typical power-angle curve of a synchronous generator. As the power angle increases, the machine output increases up to a maximum. After this point any further increase in mechanical input will exceed the stability limit, and the machine will go out of step with the power system to which it is connected. Increasing excitation increases the maximum torque angle, and decreasing excitation reduces the maximum torque angle at which the machine will pull out of step.

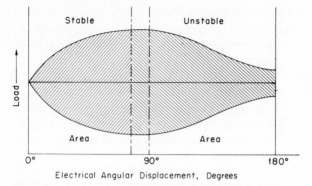

figure 2-7   Power-angle characteristics of a transmission line.

between the sending and receiving ends may exceed the stable limit. This factor is taken into consideration in determining the protection used on long lines so that the relays can sense stability limits and trip before instability occurs.

Following a power system disturbance, oscillations often occur, during which generating machines' power angles increase and decrease within a time period determined by the inertia of the machines connected to the line. In such cases it can readily be seen that the swings produced by a disturbance may cause the angular displacement of a line to exceed the stability limit on heavily loaded lines. This factor is also considered in establishing the loading limits of transmission lines.

CONTROL OF POWER FLOW WITH PHASE-SHIFTING TRANSFORMERS

When there are two or more parallel paths for power to flow between power systems or within a system, the load will divide inversely proportional to the path impedances. For example, if two circuits connect the buses of stations A and B and one circuit has an impedance of 20 Ω and the other an impedance of 10 Ω, the 10-Ω circuit will carry twice as much current as the 20-Ω circuit.

In some cases a transmission line with greater power-handling capability may be longer and have a higher impedance than a short line with low load capacity. If such lines are operated in parallel, the line with low load capability may overload before the capacity of the larger line is reached.

If voltage on the high-impedance line is increased or decreased by means of a voltage-regulating transformer installation, the load division between the lines will not be affected, but increased var flows will result, with their attendant losses.

It is well understood that power flows on a line are proportional to the angular phase displacement between the sending and receiving ends of the line. This gives a clue to how power division between parallel lines may be controlled.

Step or induction voltage regulators consist of an exciting winding and a series winding. Induction voltage regulators move the series winding with respect to the shunt exciting winding. The voltage induced in the series winding is affected by the angular position of the rotor with respect to the stator. Because of mechanical problems, induction regulators are somewhat limited in maximum size and are usually applied to distribution lines. The following discussion will be confined to step-type regulators. The exciting winding is connected across a phase, and the series winding is in the line conductor of the same phase.

Figure 2-8 shows a simplified diagram of a single-phase step-type voltage regulator.

If automatic controls are properly applied, the movable contact will move toward the tap marked R to raise voltage and toward the tap marked L to lower voltage. At the tap marked 0 the regulator will have no effect. In actual practice, a regulating transformer is considerably more complex than the diagram indicates because there is a bridging winding on the movable contacts so that they can move from step to step without interrupting the circuit and without causing a direct short circuit between taps.

If one uses basically the same device as a step-type voltage regulator and energizes the exciting winding from another phase, the voltage induced in the series winding will have an angular phase displacement from the phase in which the series winding is connected. Connection of a phase-shifting transformer is shown in Fig. 2-9 for one phase.

In the figure, only connections for one phase are shown for simplicity.

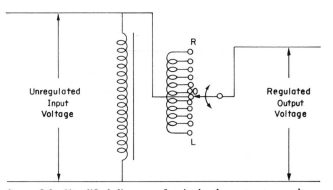

figure 2-8 Simplified diagram of a single-phase step-type voltage-regulating transformer.

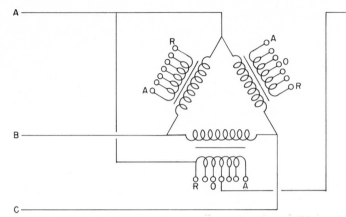

figure 2-9 Simplified diagram for one phase of a phase-shifting transformer installation.

It can be seen that the A-phase series winding is excited by the BC voltage, which is 90° out of phase with the A phase to neutral voltage. Consequently, as the tap is moved, the voltage on the phase-controlled side of the installation can be advanced to point A or retarded to point R. This can be more readily seen in the vector diagram shown in Fig. 2-10.

As indicated in Fig. 2-10, the voltages in the series windings are advanced, a condition tending to cause more power to flow in the line in which the phase-shifting transformer is installed.

By reversing the polarities in the series windings, the voltages can be retarded, with the result that the line will carry a reduced load.

By the application of proper control devices to a phase-shifting transformer installation, the power flow on a line operated in parallel with other lines can be controlled as desired.

In actual construction, although a phase-shifting transformer is similar to a voltage-regulating transformer, insulation problems are more com-

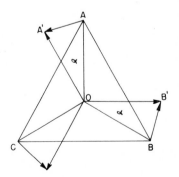

figure 2-10 Vector diagram of the effect of a phase shifter. With the phase A exciting winding energized by voltage BC, the voltage AA' is induced in the A phase series winding and will advance the A phase to neutral voltage from OA to OA' by the angle α. Similarly, the B phase voltage would be advanced by energizing the B phase series winding with the voltage CA, and the C phase by exciting the series winding with the voltage AB.

plex. Since the exciting winding is energized by a phase voltage other than that in which the series winding is connected, the insulation between the exciting winding and the series winding must be able to withstand full line-to-line voltage. Also, full-voltage bushings are required on both ends of the exciting winding, whereas in a voltage-regulating transformer one end of the winding is normally operated at neutral or ground potential.

PARALLEL OPERATION OF POWER SYSTEMS

When two or more power systems are interconnected, the tie lines behave like the transmission lines between one or more generators and a load, as discussed above. For purpose of this discussion, all the generation in each system can be considered as if it were lumped into a single machine for the system, with each system having a single equivalent load.

In transferring power between systems, power will flow from the system with greater composite leading power angle. Just as in the case of transferring load between two machines, as noted above, when it is desired to increase delivery from system A to system B, the energy input to the prime movers of generators in system A is increased, and simultaneously the energy input to prime movers of generators in system B is reduced. As a result of these changes, the power angle of system A increases and the power angle of system B decreases, with an attendant increase of power flow to system B. This is illustrated in Fig. 2-11.

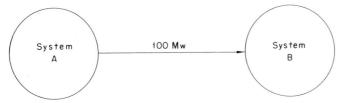

figure 2-11(a)   Energy input to system A generators = system A load + 100 MW. Energy input to system B generators = system B load − 100 MW.

figure 2-11(b)   Energy input to system A generators = system A load + 200 MW. Energy input to system B generators = system B load − 200 MW.

Following system disturbances, where one interconnected system loses generation or load, oscillations usually result because energy in the inertia of the systems is alternately given up and stored until stable conditions are reestablished. In the area giving up stored energy, the machines momentarily slow down, and in the area receiving energy, the machines momentarily speed up. This action causes fluctuating power flows on the tie lines between the systems as energy flows back and forth between them.

Usually the power swings diminish with time and disappear. In some cases, however, the time constants of the systems and of control equipment may be such that the swings increase with time until angular displacement between the terminals of the tie lines exceeds the stable limits and relay action is required to separate the systems. Computer transient stability studies are made to determine the behavior of systems under transient conditions and to determine the control action and relay protection needed to protect the systems under such conditions. Stability of power systems is discussed at greater length in Chap. 11.

PARALLELING UNITS AND SYSTEMS

When generating units are to be paralleled to a power system, or when power systems are to be paralleled, special conditions must be met in order to prevent unwanted and perhaps excessive energy flows at the time the paralleling switches are closed.

Similar requirements must be satisfied if a single generating unit is to be paralleled to an operating system or if two separated systems are to be paralleled. In paralleling two systems, however, the inertias are much greater, and more care must be taken to be sure that proper conditions exist before the paralleling switch is closed.

In paralleling machines or systems, four conditions must be satisfied:

1. Phase rotations must be the same.
2. The electrical speed of the machine or system that is being paralleled must be the same as that of the system to which the parallel is being made.
3. The machine and system or the two systems must be in phase; that is, there must be little or no angular difference between the corresponding phases.
4. The voltage of the machine or system that is being paralleled should match the voltage of the system to which the parallel is being made at the point of the parallel.

The phase rotation is not usually a problem to the power system operator, since the rotations are determined by test, and once connections are properly made the rotation is fixed and can be changed only by altering phase connections of the machine or the bus at the station involved. However, after equipment has been overhauled, or whenever connections are disturbed, the system operator should make sure that proper phase-rotation checks have been made before attempting to place the equipment in service. For the rest of this discussion it will be assumed that rotations are correct.

When a generating unit is to be paralleled to a power system, the inertia of the machine is usually much less than that of the system. Furthermore, the speed and voltage of the machine can readily be varied until the system speed and voltage are matched.

To enable an operator to observe the conditions for synchronism, synchroscopes, synchronizing lights, and voltmeters for both the incoming machine and the system are provided. A synchroscope is a device which produces a revolving field proportional to the speed difference between the incoming machine and the system. It is constructed so that the angular phase difference is indicated on the synchroscope dial. Consequently, when a machine is in phase with the system to which it is to be paralleled, or when two systems are in phase, the synchroscope needle will be stationary with zero angular displacement. Synchronizing lights are connected between potential transformers of the incoming machine or system and the running system and indicate the voltage difference. Such lamps can be connected to be either dark or bright when an in-phase condition exists. Simplified connections of a synchroscope and synchronizing lights are shown in Fig. 2-12.

If the machine shown in Fig. 2-12 is running at less than synchronous speed, even though the synchronizing switch is closed at exactly the instant that the synchroscope shows the machine to be in phase with the system, there will be a flow of energy from the system to the incoming machine to accelerate it to synchronous speed. If the speed difference is significant, damage to the machine could result, since heavy currents will flow in the machine windings to develop a motor torque to produce the required acceleration. The more nearly the machine speed is matched to the system speed, the less the disturbance when the synchronizing switch is closed, assuming, of course, that synchronism is indicated on the synchroscope.

If the incoming machine is running somewhat faster than synchronous speed, and the synchronizing switch is closed when the synchroscope indicates an in-phase condition, there will be a flow of energy from the machine to the system to retard the machine to synchronous speed. Again the amount of energy flow is proportional to the speed difference.

If the incoming machine is running at synchronous speed, but a phase difference is indicated on the synchroscope at the time the switch is

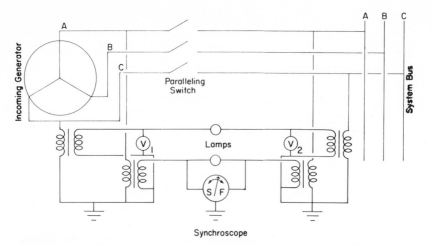

figure 2-12   Simplified diagram of synchronizing facilities. The machine and bus volt-ages are reduced to instrument levels by potential transformers. The voltmeters V_1 and V_2 serve to indicate the voltage of the incoming machine and the system bus. The lamps, as connected in the diagram, will be dark when an in-phase condition exists and will flicker at a rate equal to the frequency difference if the machine speed is other than synchronous. The synchroscope will indicate difference in phase and will rotate in a direction determined by whether the incoming machine is above or below syn-chronous speed.

closed, heavy currents will flow to either accelerate or retard the machine as may be required to bring it into time phase with the running system.

The objective in paralleling a machine to a system or in paralleling systems is to match speed and phase position so that there will be little or no transfer of energy between the incoming and running equipment at the time the paralleling switch is closed.

The reason for circulation of current between a machine and a system or between two systems when the synchronizing switch is closed with a phase difference existing is illustrated by the vector diagrams of Fig. 2-13.

It should be pointed out that if voltages are not matched at the time the synchronizing switch is closed, there will be a var flow from the system to the incoming machine if the system voltage is high, or from the machine to the system if the machine voltage is high. Although it is desirable to have voltages matched, it is much more important to match speed and phase position.

Special problems occur when paralleling systems together, since the inertias involved may be many times greater than those involved in paralleling a single machine to a system. If speed and phase position are not very carefully matched, exceedingly high currents can result, with possibly damaging results. Angular phase displacements that would be tolerable when paralleling a machine to a system may cause relay operations to occur or damage to equipment when two systems are paralleled. Power will flow from the leading system to the lagging system and will tend to accelerate the lagging system and decelerate the leading system. Because of the great difference in inertias, either the current flows will be much greater or the excessive flows will persist longer than when paralleling a single unit with the system. In either event, the potential for equipment damage or system breakup exists, so that precise speed and phase-position adjustment must be made to parallel systems successfully.

Another problem in paralleling systems together is that the point of interconnection is usually remote from points of generation. Consequently, system speed adjustments must be made by telephonic orders to generating plants or by remote control equipment. Because the frequencies of both systems may be very close to 60 Hz, it may take considerably longer to obtain the correct synchronous phase position than when a single unit is being paralleled. Normally paralleling is done at attended stations where synchroscope indication is available. If the paralleling point is not attended, remote synchronism indication can be provided to the system operator.

Automatic synchronizing equipment of the electromechanical type is not usually precise enough to permit automatic synchronizing between large systems; however, there have been some applications of automatic synchronizing equipment for this purpose. Electronic automatic synchronizing equipment has been developed; this can be set much more

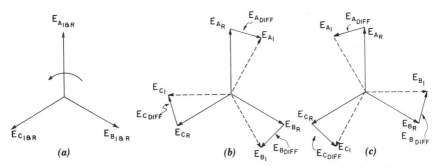

figure 2-13 Vector diagrams of voltages for conditions of an incoming machine to a running system. *(a)* In phase. The incoming and running voltages are equal and in phase and are the correct condition for paralleling. If the synchronizing switch is closed under this condition, no disturbance will result. *(b)* Incoming machine lagging. If the incoming machine is lagging, the running system and the resulting difference voltages will cause current to flow from the system to accelerate the incoming machine and bring it into the correct phase position. *(c)* Incoming machine leading. If the incoming machine is leading the running system, the difference voltages will cause current to flow from the machine to the system to retard it to the correct phase position.

precisely than electromechanical equipment, and should be capable of automatically synchronizing systems. It should be pointed out that automatic reclosers on synchrocheck devices should be out of service when paralleling large systems. With the very small speed differences that may exist between the systems, the time delay of the synchrocheck relay may permit closure to occur within the phase-angle limits of the synchrocheck relay (usually $\pm$ 30°), but with a phase angle between the systems that is unsatisfactory for successful paralleling.

Another problem that sometimes requires the system operator's attention is the closing of a parallel tie on a long transmission loop. Such a situation is illustrated in Fig. 2-14.

In cases such as that illustrated in Fig. 2-14, if the loop is closed with a large angle across the closure point, the angle will be immediately reduced to zero, and the power angles of the machines close to the closure point may require drastic adjustment, which could result in damage to them.

The angle across a closure point may be adjusted by altering generation, particularly by the machines near the closure point. In such cases generation of the machines on the leading side is reduced and that of machines on the lagging side is increased until the angle is reduced to a point at which closure can be made successfully and safely.

It should be pointed out that such adjustments will affect loop power flow on intervening systems, but this can be readjusted to some extent after the loop is closed.

Another method of effecting a loop closure is to sectionalize the loop and make the final closure at a location at which the angle across the open point is a minimum.

In general, when synchronizing machines or paralleling systems, a minimum of energy transfer is desired at the time of closing the paralleling switch. After the switch is successfully closed, energy transfer

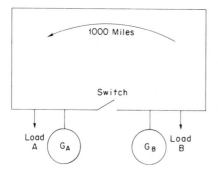

figure 2-14 Diagram of a long loop. A power angle will exist across the switch, which will be affected by the generation and loads at A and B. A maximum power angle will exist if either generator A or B is carrying the total load A and B. A minimum or zero angle will exist if the generation at A equals the load at A and the generation at B equals the load at B.

will result as determined by the prime mover input to a machine or from the system with the leading power angle. With the increased amount of interconnection between systems, the paralleling of systems is becoming of greater interest to system operators, and as systems become larger the requirements for successful paralleling become more stringent.

PROBLEMS

1. The usual distinction between transmission lines and distribution lines is
    (a) The amount of current carried by the lines
    (b) The operating voltage
    (c) The size of the conductors
2. If the load on an isolated generator is increased without increasing the power input to the prime mover:
    (a) The generator will slow down.
    (b) The generator will speed up.
    (c) The generator voltage will increase.
    (d) The generator field current will increase.
3. When two dc generators are operated in parallel and the field current on one of them is increased, it will
    (a) Take less load
    (b) Speed up
    (c) Take a larger share of the load
    (d) Cause the generator to overheat
4. When two ac generators are operated in parallel, and the field current on one of the generators is increased, it will
    (a) Take a larger share of the load
    (b) Speed up
    (c) Cause a flow of reactive between the two machines
    (d) Take a smaller share of the load
5. Changes in load division between ac generators operating in parallel are accomplished by
    (a) Adjusting the generator voltage regulators
    (b) Changing the energy input to the prime movers of the generators
    (c) Lowering the system frequency
    (d) Increasing the system frequency
6. When the energy input to the prime mover of a synchronous ac generator operating in parallel with other ac generators is increased, the rotor of the generator will
    (a) Increase in average speed
    (b) Retard with respect to the stator revolving field
    (c) Advance with respect to the stator revolving field
7. An ac generator is operating with 100-A field current. If the field current is increased to 125A with the same electrical load on the machines, it will
    (a) Be less apt to go out of synchronism
    (b) Be more apt to go out of synchronism
    (c) Operate at a new torque angle
    (d) Overheat
8. Phase shift occurs between the sending and receiving ends of ac transmission lines as a result of the
    (a) Reactances of the lines
    (b) Resistances of the lines
    (c) Voltage at which the lines operate
    (d) Conductor size

9. A transmission line is operating, for a given load, with a phase shift of 80° between the sending and receiving ends of the line. If the load is suddenly increased, the phase displacement will
   (a) Decrease
   (b) Increase
   (c) Not be affected
   (d) Become zero

10. When power is transferred between two power systems, power will flow from the system with
    (a) The greater leading power angle
    (b) The lesser leading power angle
    (c) The higher voltage level

11. Two power systems, A and B, are operating in parallel. If system A increases its generation to deliver 100 MW to system B, what will be the effect if system B does not simultaneously reduce its generation?
    (a) Frequency will decrease.
    (b) Frequency will not change.
    (c) Frequency will increase.
    (d) System B voltage will rise.

12. When a phase-shifting transformer's taps are moved in such a direction as to advance the phase position:
    (a) Var flows will increase.
    (b) There will be an increase of power flow in the line.
    (c) Var flows will decrease.
    (d) Voltage will be increased.

13. When paralleling a generating unit to a power system, if the synchroscope indicates that the incoming machine is running slower than the system:
    (a) The field current should be increased.
    (b) The power input to the prime mover should be increased slightly.
    (c) The power input to the prime mover should be decreased slightly.
    (d) The synchronizing switch should be closed just after it indicates zero phase angle difference.

14. Power system A is to be paralleled with power system B. The synchroscope across the paralleling switch is stationary, with system A lagging system B by 90°.
    (a) System A should increase speed slightly and have the synchronizing switch closed when the phase difference between the systems is near zero.
    (b) Either system should adjust speed slightly and return to normal speed when the synchroscope shows zero phase angle between the systems.
    (c) System A should increase its voltage.
    (d) Systems A and B should both increase speed slightly.

15. It is desired to close a long loop like that shown in Fig. 2-14. The synchroscope indicates that the system A side of the closure switch is 60° ahead of the system B side. In order to reduce the angle across the switch:
    (a) System A should increase generation and system B reduce generation.
    (b) System B should increase generation and no action be taken by system A.
    (c) System A should reduce generation and system B increase generation.
    (d) System B should increase its voltage.

3

Var Flows

In Chap. 1 it was shown that when current and voltage are not in phase, the product of these two quantities is referred to as voltamperes. The power in the circuit is the product of the current times the voltage times the cosine of the angle between the current and voltage in the single-phase case, or

$$\text{Power} = EI \cos \theta \qquad (\cos \theta = \text{power factor})$$

as shown in Fig. 1-6a, or

$$\text{Power} = \sqrt{3}\, EI \cos \theta$$

in the three-phase case.

The reactive voltamperes are:

Single-phase case: $\text{var} = EI \sin \theta$ (Fig. 1-6b)
Three-phase case: $\text{var} = \sqrt{3}\, EI \sin \theta$

LOSSES DUE TO VAR

Var (voltamperes reactive) in an ac electric system always cause an increase in current, which results in increased losses. This will be shown as follows.

All transmission and distribution lines contain resistance, inductance, and capacitance. The current flowing through the resistance is in phase, and the product of the resistance voltage drop and the current represents a power loss in the conductor.

$$\text{Power loss (watts)} = \text{voltage drop } E \times I \text{ (line amperes)}$$

as $$E = I \times R \qquad \text{from Ohm's Law}$$

or $$\text{Power loss} = I \times R \times I$$
$$= I^2R \qquad \text{watts}$$

From this relationship it can readily be seen that doubling the current in a circuit results in four times the power loss. In a circuit with a power factor of 0.5, the current would be twice that which would flow if the power factor was 1. At heavy loads on a line, the losses due to var flows can become very significant. In addition, as a result of the increased current in a circuit with var flows, the voltage drop due to the line resistance is greater than it would be at unity power factor.

In a dc transmission line the voltage at the receiving end of the line is always less than the voltage at the sending end of the line by an amount determined by the line current and the line resistance. This relationship is as follows: The receiving-end voltage equals the sending-end voltage less the voltage drop in the line, which is equal to the line current times the line resistance, or, mathematically,

$$E_R = E_S - IR_L$$

where $$E_R = \text{receiving-end voltage}$$
$$E_S = \text{sending-end voltage}$$
$$I = \text{line current}$$
$$R_L = \text{line resistance}$$

The ac case is much more complex. The inductance of a line is distributed throughout its length, and capacitance exists between conductors and also between conductors and ground. This is also distributed throughout the length of the line. As a result, in a line of appreciable length, even with a unity-power-factor load, a capacitive var input is required to supply the charging current of the line. The amount of charging current is determined by the capacitive reactance of the line and always leads the voltage in phase position. However, as current flows along the line from the sending to the receiving end, it encounters inductive reactance.

At light loads, the shunt capacitive current may exceed the load current, and the line will operate at a leading power factor from the sending end. Voltage drops or rises will result, made up of the current times the capacitive reactance, the current times the inductive reactance, and the current times the line resistance. Since these drops add as vectors, the result can be that the voltage at the receiving (load) end of the line is higher than that at the sending end. As load current increases, the voltage drop through the series inductive reactance increases while the capacitive current remains constant. Consequently, at some load the capacitive and inductive components are equal and, for any increase in load, the inductive reactance drop will exceed the capacitive reactance effect. Under these conditions the line drop will exceed that which would exist if only resistance were present.

As a consequence of the line inductance and capacitance, a *transmission line always requires a var input,* which may be either leading in light load conditions or lagging in heavily loaded conditions.

At the receiving end of a line, the power factor is determined entirely by the power factor of the loads connected to the line, including the station transformers, which are inductive and require lagging var. If the load is anything other than unity power factor (purely resistive), additional var supply will be required. With heavy loads and at low power factors, the var requirements can equal or exceed the load (watt) requirements.

VAR COMPENSATION

Various methods are used to supply var needed in a power system. Synchronous condensers or generators can supply either leading (capacitive) or lagging (inductive) var. Static capacitors can be connected in parallel across loads supplying leading var to correct for lagging power factor of motors or other inductive equipment and on station buses to compensate for lagging var requirements of station transformers and of the lines from the station. Capacitors are also installed on distribution lines to compensate for customer var requirements. Many such installations are automatically switched so that the capacitors are connected to the line only when needed.

Capacitors are sometimes connected in series in a line. In such cases the current through the series capacitors varies as the load current varies, and the voltage rise across the series capacitor bank (IX_C) will offset the voltage drop resulting from the series inductive reactance (IX_L) of the line. Installations of this type are used frequently on long extra-high-voltage transmission lines. In addition to reducing voltage regulation of a line, the use of series capacitors can materially increase the line's stability by reducing the angular phase displacement between the sending and receiving ends.

As mentioned above, during light loading conditions with a long line, the receiving-end voltage can exceed the sending-end voltage and may, in some cases, become excessive.[1] To compensate for this condition, shunt reactors are installed at station buses or on the tertiary windings of transformer banks. By drawing lagging currents, shunt reactors directly offset the leading currents due to line charging. By convention, when reactive (var) flow is from the station to the system, the flow is considered to be positive (+), and when the var flow is from the system to the station bus, it is considered to be negative (−).

Shunt and series capacitors and shunt reactors are switched into circuits in increments (blocks) as needed to satisfy the requirements of the system under existing load conditions. Switching of such var sources can be either manual or automatic. If such devices are left in service at all times, they can worsen rather than help voltage conditions.

Synchronous condensers and generators, as has been noted, can supply either leading or lagging var merely by field-current adjustments. Synchronous condensers are synchronous motors without a connected mechanical load. These machines can usually go to full rating in providing leading var (overexcited operation) and approximately 50 to 80 percent of rating in absorbing lagging var (underexcited operation). Although there are many synchronous condensers presently installed, few new ones are being installed because equivalent var sources can be purchased at less cost in static capacitors and shunt reactors. Moreover, no mechanical maintenance of the static devices is required.

Static var compensators have been developed that can, in many cases, replace synchronous condensers. Capacitive and inductive elements are connected in parallel, with thyristors in one or both legs. The firing point of the thyristor gates can be controlled so that more or less inductive or capacitive current will flow. By such means the static compensator can act as a capacitive var source or as an inductive var load, and can act as the equivalent of a synchronous condenser.

GENERATORS AS VAR SOURCES

Probably the greatest source of controllable var available to a power system operator is generating equipment. Many machines are rated at

---

[1]This topic is discussed in greater detail in Chap. 12.

something other than unity power factor, for example, 0.8. This means that the MVA rating of a 100-MW generator is 125 MVA. Assuming that there are no other limitations, such as maximum or minimum bus voltages, the machine would be able to supply approximately 75 Mvar at full load without exceeding its MVA rating. When a generator is carrying electrical load, the ability to operate it "leading" to absorb lagging var (underexcited) may be limited because, with reduced excitation, the power angle is increased and the machine may pull out of step. The amount that a generator can operate in the lead is determined to a great extent by how fast the field voltage control can respond. Modern machines with electronic voltage-control systems, amplidynes, etc., can operate safely in the "lead" with low excitation to a much greater degree than is possible with machines equipped with rheostatic or other relatively slow field-controlled systems.

Before leaving the subject of var supply from generating equipment, it might be well to tabulate the var available at various power factors from a generator at 100 percent MVA loading:

% Power Factor	MW, %	Mvar, %
100	100	0
95	95	30
90	90	43
85	85	53
80	80	60
75	75	66
70	70	70
65	65	76

VAR FLOWS DUE TO UNBALANCED VOLTAGES

Another factor affecting var flows in a power system is the transformer winding ratios at substations of an interconnected system. Proper selection of transformer taps can materially reduce unwanted var flows. For example, assume two substations supplied from the same high-voltage transmission line and operated in parallel on the secondary side. If the transformer taps at one station are set for a higher secondary voltage (lower turns ratio) than those at the other station, var will flow from the station with the higher voltage in an amount that will cause sufficient impedance drop in the connecting line so that connecting line and bus voltages will be equal at the lower-voltage station. This is illustrated in Fig. 3-1.

It should be pointed out that control of the flow of var is generally a local problem, in contrast to control of the flow of power, which is a

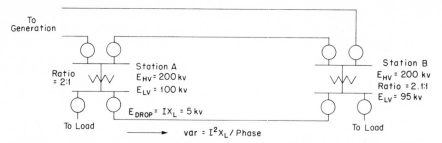

figure 3-1 Diagram showing var flow in one phase due to unequal voltages on substation buses connected in parallel. A reactive-current component will flow from station *A* (station with higher voltage) to station *B*, limited by the circuit reactance, in an amount that will cause an IX_L voltage drop equal to the voltage difference between the two stations. This current squared times the line reactance will be the var flow resulting from the voltage difference. In a three-phase system the total var would be $3I^2X_L$.

system problem. Because of the numerous interacting factors, including transformer taps, capacitor installations, reactors, generator voltage control, and transmission-line reactive generation, a mismatch of var requirement in one area can result in high or low voltage in that area but have little or no effect on remote portions of the system. Because of the diverse problems, automatic economic control of voltage and var, while possible, can be expensive from a control-equipment standpoint.

SUMMARY

The following will summarize the previous discussion on var flows:

1. Var are required in an ac power system due to the capacitive and inductive reactances of power system lines and equipment and customer loads.

2. Var in excess of those required to satisfy circuit requirements represent an incremental power loss in the system.

3. Var sources can be used to limit var flows due to low-power-factor loads and to control voltages on station buses.

4. Generation equipment, properly operated, can be used to supply much of the var requirements of a system.

5. Proper selection of transformer taps at interconnected stations will minimize var flows between the stations.

PROBLEMS

1. Power in a single-phase ac circuit is the product of
   (a) Voltage times the current in the circuit
   (b) Voltage times current times the cosine of the angle between the voltage and the current

(c) Voltage times current times the sine of the angle between the voltage and the current

2. When var flow in a circuit, losses in the circuit are
   (a) Reduced
   (b) Not changed
   (c) Increased
3. In a lightly loaded circuit of such a length that the capacitive reactance is appreciable, the receiving-end voltage
   (a) Is always less than the sending-end voltage
   (b) May exceed the sending-end voltage
   (c) Is always equal to the sending-end voltage
4. Why does an ac transmission line require a var input?
5. On a long, high-voltage transmission line under heavy load conditions, var compensation can be provided by installing
   (a) Series inductive reactors
   (b) Series capacitors
   (c) Shunt inductive reactors
6. When synchronous generators or condensers are used to provide var, leading var are produced by
   (a) Increasing the field current
   (b) Reducing the field current
   (c) Increasing the speed of the machines
7. With a 100-MVA generator operating at 85 percent power factor lagging,
   (a) How many Mvar will be produced?
   (b) To what megawatt load should the machine be limited so that its MVA rating will not be exceeded?
8. Var are characterized by the fact that they always flow
   (a) From points of low voltage to high voltage
   (b) Without effect by the voltages at the line terminals
   (c) From points of high voltage to low voltage
9. In the circuit shown in Fig. 3-1, assume that there are 100 A flowing in the 100-kV circuit between stations A and B. Determine
   (a) The inductive reactance per phase of the circuit
   (b) The total kvar input to the circuit at station A

4

Economic Operation of Power Systems

Successful operation of power systems requires attention to safety of personnel and equipment and the provision of service to utility customers without interruption and at the lowest feasible cost. The problem of providing low-cost electric energy is affected by such items as efficiencies of power-generating equipment, cost of installation, and fuel costs for thermal-electric plants. Factors involved in the cost of producing energy can be divided into two categories: those that are fixed and those that are variable.

FIXED COSTS

Fixed costs include capital investment, interest charges on borrowed money, labor, taxes, and other expenses that continue irrespective of the load on the power system. Persons responsible for the operation of a power system have little control over these costs.

VARIABLE COSTS

Variable costs are those costs which are affected by the loading of generating units of different fuel or water rates, control of losses caused by reactive flows, the combination of hydro and thermal generation to meet daily load requirements, and purchase or sale of power. These costs are materially controlled by power system operators. This section will discuss factors in power system operation that can be controlled and methods used to ensure that power generated to carry the power system load is always produced in such a way that minimum costs will result. The savings that can be achieved by proper operation of power resources can be very significant; they may amount to several thousand dollars a day on large power systems.

Many power systems have several alternative sources for electric energy, such as conventional steam-electric plants, nuclear plants, hydro, geothermal, gas turbine, and outside sources from which power may be purchased. Solar, wind power, and fuel cells are also alternative sources of electric power. Considerable work is being done to make the energy cost of such sources competitive with that of more conventional sources of energy. Normally the electrical capacities of the units of such sources are much less than the capacities available from conventional thermal and large hydro units. The continuing problem is to determine at all times the combination of sources, and loads on these sources, which will result in minimum overall production cost.

Fuel supplies for thermal plants can be natural gas, oil, nuclear sources, or coal, with varying costs for each. Furthermore, the load on a power system is continually changing. For this reason the economic supply problem must be reviewed frequently, and, if necessary, load allocations on the various power sources readjusted so that deviations from the most economic operation will be held to a minimum.

Water supplies for hydro generation can have different values from time to time, and the use of hydro power must be integrated into the system power supply so that the lowest overall costs result.

Pumped storage is a special type of hydro power in which water is pumped to an upper reservoir during "off peak" hours when thermal generation costs are at a minimum. The pumped water is released during "peak" hours to generate hydro energy, and thus replace thermal generation when fuel costs would be high. It should be pointed out that pumped-storage generation requires more energy for pumping than is recovered during the generation cycle. However, the value of the power generated during the peak load periods will normally more than offset the cost of the thermally generated power used for pumping.

Power exchanges between interconnected systems can also be used to advantage in minimizing fuel costs when there are significant differences in generation costs of interconnected power systems.

Because the characteristics of the various types of power supplies differ, each type will be considered individually. An effort will then be made to develop the basis for the procedures used to attain overall economy when the different types of power sources are used simultaneously. Consideration will be given first to conventional thermal plants.

THERMAL POWER PLANT EFFICIENCIES

It is a well-established physical principle that as the differences between the temperature and pressure inputs and outputs of a heat-operated device, such as a steam turbine, are increased, more mechanical power output will be developed for the same amount of heat energy input. This is the basic reason for the ever-increasing pressures and temperatures in modern steam-electric generating units. The overall efficiency of thermal units is determined by measuring the heat input and the electric energy output and expressing the results as a ratio at various loads. Curves called input-output curves can be drawn from the results of such tests, with energy input expressed in Btu (British thermal units), equivalent barrels of oil, or Mcf of gas per hour, and the output as kilowatts or megawatts. A characteristic of these curves is that fuel input is increased as electrical output is increased, but not necessarily linearly. Such a curve is shown in Fig. 4-1.

Curves of this type are developed for each generating unit involved. From them it can readily be seen that efficient units will develop a given amount of power with less fuel input than will be needed by units of

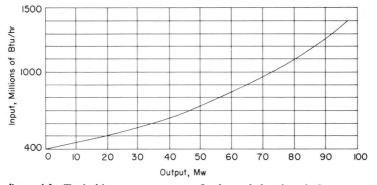

figure 4-1   Typical input-output curve of a thermal-electric unit. Input can be expressed in Btu, equivalent barrels of oil, or Mcf of gas. (An equivalent barrel is 6,250,000 Btu, and an Mcf of gas may vary depending on the source but is usually approximately 1,000,000 Btu.)

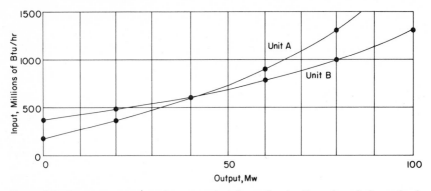

figure 4-2 Input-output curves for two typical thermal units. Even though the no-load fuel for unit *B* is greater than that of unit *A*, at loads above 40 MW the heat input for unit *B* is less than that for unit *A*.

lower efficiency. The first and obvious conclusion might be to load the efficient units before loading the less efficient units. This would, of course, be a better solution than loading the low-efficiency units first, but the desired solution is to load the available units so as to develop the required power at the least possible cost. Techniques to solve this problem have been developed and are in general application throughout the industry.

Consider two 100-MW thermal units *A* and *B* with input-output curves as shown in Fig. 4-2.

INCREMENTAL RATES

It can be shown mathematically that minimum fuel input for any given total load of the two machines will occur when they are operated at equal incremental heat rates. Because fuel has a cost, such as cents per Btu, dollars per equivalent barrel, or cents per Mcf, the above statement can be modified to say that *the minimum cost will occur when the incremental costs are equal.*

The term "incremental" merely means a small increase. Of course, the smaller the increment (increase), the more precise the determination of incremental change. An incremental rate is defined as the slope of a curve from one point to another. Examples of the determination of incremental rates are given in Fig. 4-3.

An inspection of Fig. 4-3 gives a clue to an easy method of determining incremental rates. If it is assumed that the curve is made up of straight-line segments between the numbered points, then we have already calculated the incremental rates (slopes) for points *B* and *D*. The slope at

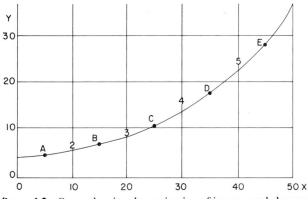

figure 4-3 Curve showing determination of incremental changes. For a small distance along a curve it can be considered to be a straight line. The increments on the x axis in this case are from 10 to 20 between points 2 and 3 and from 30 to 40 between points 4 and 5. When following the curve from 10 to 20 on the x axis, it goes from 6 to 8 on the y axis. The slope of this portion of the curve is a ratio of the differences, or $(8 - 6) \div (20 - 10)$, which equals $2/10 = 0.2$. When following the curve from 30 to 40 on the x axis, it goes from 14 to 22 on the y axis. The slope then is $(22 - 14) \div (40 - 30)$, which equals $8/10 = 0.8$. This method, using smaller and smaller increments until they approach zero, is the basis for differential calculus. For practical purposes in determining incremental costs, little error is introduced by using reasonably small increments other than those approaching zero.

A would be $(6 - 4) \div (10 - 0)$ or $2/10 = 0.2$. At point C the slope would be $(13 - 8) \div (30 - 20)$ or $5/10 = 0.5$.

At point E the slope would be $(36 - 22) \div (50 - 40) = 14/10 = 1.4$. An incremental-rate curve is developed by plotting the points determined by the above calculations. This is shown in Fig. 4-4.

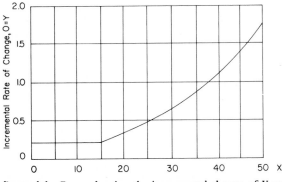

figure 4-4 Curve showing the incremental change of Y as X increases for the curve shown in Fig. 4-3.

TABLE 4-1 Procedure for Determining Load Allocation for Two Units Whose Curves Are Shown in Fig. 4-2*

(1)	(2)	(3)	(4)	(5)
			Charge for	Incremental rate,
Load,	Million	Dollars/h	load increment,	dollars/MWh
MW	Btu/h	(2) × \$3.50	dollars/h	(4) ÷ (1)
		Unit A		
0	200	700		
			350	17.50
20	300	1050		
			525	26.30
40	450	1575		
			700	35.00
60	650	2275		
			1045	52.20
80	950	3320		
			1930	96.50
100	1500	5250		
		Unit B		
0	250	875		
			350	17.50
20	350	1225		
			350	17.50
40	450	1575		
			525	26.20
60	600	2100		
			700	35.00
80	800	2800		
			875	43.80
100	1050	3675		

*The fuel costs shown in this tabulation have been increased by a factor of 10 to more realistically reflect present costs, which have drastically increased since the first edition was prepared.

ECONOMIC LOADING OF GENERATING UNITS

When the basis for determining incremental-rate curves has been developed, this method can be used to determine how to operate electric generating units for minimum production cost. A procedure for determining load allocation for minimum fuel cost between the two units whose input-output curves are shown in Fig. 4-2 is shown in Table 4-1. Since we are primarily interested in cost, the fuel rates will be converted to dollars per hour for various loads. The incremental rate in dollars per megawatt-hour for various loads is determined. A fuel price of \$3.50 per million Btu is assumed.[1]

The incremental-rate information developed in Table 4-1 can be plotted as curves for both machines, Figs. 4-5 and 4-6. If 1-h periods are considered, the vertical scale will be incremental cost in dollars per megawatthour, with megawatts as the horizontal scale, as shown in Fig. 4-6.

[1]Fuel costs have increased by a factor of at least 10 since the early 1970s. This has significantly increased the cost of electric power for all consumers.

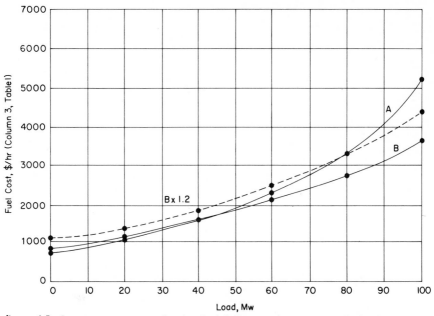

figure 4-5 Input-output curves showing fuel dollars per hour versus MW load.

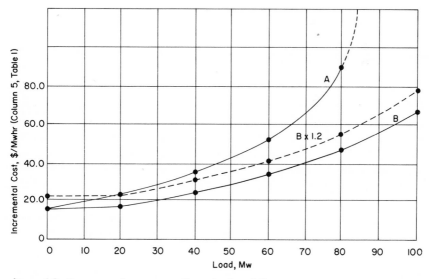

figure 4-6 Incremental cost curves for units *A* and *B*.

From the information shown in Figs. 4-5 and 4-6 it should be possible to determine the proper division of load between the two machines to result in minimum fuel cost. Assume a total load of 100 MW to be carried by the two units. Various combinations of loading can be made, but the objective is to carry the load with minimum cost. The tabulation shown in Table 4-2 was developed by taking points on the incremental fuel-cost curves of Fig. 4-6 to match various load conditions and fuel cost per hour from the input-output curves shown in Fig. 4-5, where fuel cost in dollars per hour is plotted against loads in megawatts.

From Table 4-2 it can be seen that the minimum fuel cost occurs when unit A is loaded to 40 MW and unit B to 60 MW with incremental costs of $30 MWh for each machine. If desired, tabulations similar to that in Table 4-2 can be set up for other fuel costs and the minimum cost (fuel rates) determined as further proof of the principle of loading machines for equal incremental costs.

It is obvious that when many units are involved, a manual solution of the economic loading problem is impractical, because many load changes would be needed while solving the problem for only one situation. Various devices have been developed to help solve the economic loading problem rapidly where many generating units are involved.

Probably the simplest device for allocating load on an incremental basis is the economic loading slide rule. These devices were used to a considerable extent prior to the advent of digital computers with economic loading programs.

The slide rules made use of sliding elements showing unit loading on

TABLE 4-2

Unit A			Unit B			
Load, MW	Incr. fuel cost, dollars/MWh	Fuel cost, dollars/h	Load, MW	Incr. fuel cost, dollars/MWh	Fuel cost, dollars/h	Total fuel cost, dollars/h
0	15.50	730.00	100	48.50	3650.00	4380.00
10	17.50	830.00	90	43.80	3220.00	4050.00
20	21.00	1010.00	80	39.00	2800.00	3810.00
30	26.30	1250.00	70	35.00	2450.00	3700.00
40	30.00	1575.00	60	30.00	2100.00	3670.00
50	35.00	1870.00	50	26.50	1825.00	3695.00
60	43.00	2250.00	40	21.00	1575.00	3820.00
70	52.50	2750.00	30	17.50	1400.00	4150.00
80	66.00	3320.00	20	17.50	1228.00	4540.00
90	96.50	4100.00	10	17.50	1050.00	5150.00
100	—	5200.00	0	17.50	900.00	6100.00

logarithmic scales, and a straightedge that could be adjusted in position. The sliding elements could be set to the unit fuel cost, and by moving the straightedge to the appropriate position, unit loadings could be determined for minimum overall fuel cost.

COMPUTERS FOR ECONOMIC LOADING

Digital computers are almost universally used for economic loading of generating units. Although there may still be some analog load-frequency-control (LFC) systems in use, they are rapidly being supplanted by digital systems.

The particular advantage of computers is that they can continuously monitor the system loading conditions, determine the most economical allocation of generation between units, and send control impulses to load the units to the desired values. Computer control, properly applied, can approach an almost exact allocation of unit loadings for minimum fuel cost. In applying computers to online economic loading problems, the unit input-output curves and incremental-fuel-rate curves are stored in the computer, which goes through a process similar to that followed in Table 4-2 to calculate the desired machine loadings. Because of the tremendous speed at which computations can be made in a digital computer, it can solve economic loading problems in very short time intervals and simultaneously carry out other system-control functions.

Automatic generation control (AGC) is commonly included in supervisory control and data acquisition system (SCADA) installations. Such systems are described in Chap. 8.

EFFECTS OF VARYING FUEL COSTS

Before leaving the problem of incremental loading of thermal plants, the matter of varying fuel costs should be mentioned. The shapes of the input-output and incremental-fuel-rate curves are not changed by different fuels or by changes in the cost of the same fuel. Consequently, if the incremental curves are plotted with incremental cost as the vertical scale, the ratio of the cost of the fuel being burned to the cost of the fuel for which the curves were drawn can be used as a multiplying factor. This factor is employed to correct for fuel-cost changes for any or all of the units. By this means it is possible to solve the economic loading problem under all conditions of fuel cost.

There is a further complication in accounting for losses due to transmission of power from generation to load. This will be discussed in more detail later. It will suffice for the moment to state that transmission losses can be, and are, evaluated and their effect used as a multiplier on the incremental fuel cost of each unit to provide a means of obtaining actual overall economy, including transmission losses.

NUCLEAR GENERATION

The above discussion has been confined to the loading of conventional fossil-fueled thermal plants. Nuclear plants present a special problem. Owing to various factors it is usually desirable to operate nuclear units at base load and at full output at all times. Consequently, it is not necessary to consider them as an incremental source to the system. However, in supplying a portion of the energy that otherwise would have to be supplied by conventional thermal units, nuclear units cause these other thermal units to operate at a reduced incremental cost.

GEOTHERMAL GENERATION

As a result of escalating fuel costs in recent years, there has been greatly increased interest in the use of geothermal energy for electric power generation. This source can be used only where there are sources of natural steam or hot water that can be economically developed. The largest such development is the Geysers of the Pacific Gas and Electric Company in California.

Like nuclear installations, the units at geothermal plants are normally operated as base-load units and not on an incremental basis.

SOLAR AND WIND POWER GENERATION

There has been progress in the development of solar and wind power in recent years. These developments are also due, to a considerable extent, to the increasing costs of fossil fuels.

At present such developments are of relatively small capacity, and because of the variable nature of sun and wind, neither source is adaptable to incremental loading techniques. Power from these sources is developed when sun and wind conditions are favorable, and to the extent available it will reduce the use of normal fossil fuels.

COORDINATION OF HYDRO AND THERMAL GENERATION

The operation of hydro units in a system in which both hydro and thermal generation are used presents an extension of the economic loading problem. There are many conditions connected with hydro operations, such as uncontrolled flows and required releases of water for irrigation or flood control, which take away from the system operator some of the alternatives that might be available if the water could be used entirely as desired for the benefit of power production. However, if a value can be placed on water, usually in dollars per acre-foot, hydro units can be operated incrementally along with thermal units for overall economic operation of the system.

Of course, the value of water changes from time to time, being lowered during periods of high flow, during and immediately following storms, and increased during periods when flows are low or when reservoirs are being drafted at controlled rates of flow. Since each acre-foot of water through a hydro plant will develop a definite amount of energy, depending on the head of the plant, water is equivalent to fuel such as gas or oil for power-producing purposes.

Procedures for integrating the operation of hydro and thermal generation on a system for minimum cost of generation have been developed and are in use. This procedure is called hydrothermal coordination.

Basically, in a hydrothermal-coordination program, input-output curves for each hydro unit are developed, showing acre-feet per hour plotted against load in megawatts. From the input-output curves, incremental-water-rate curves showing the incremental water rate in acre-feet per megawatthour plotted against load in megawatts can be developed by exactly the same method used for thermal plants.

An arbitrary price in dollars per acre-foot is placed on the water for each plant. If it is desired to use more water, the price is reduced, and if less water is to be used, the water price is increased. By proper selection of water prices, exactly the desired amount of water will be used during any desired time period. The hydro plants then will follow incremental loading requirements of the system and help achieve the desired result of overall minimum fuel cost.

The water value in hydrothermal-coordination programs is usually denoted by the Greek letter gamma (γ) to distinguish it from the thermal unit and system fuel cost, which is designated by the Greek letter lambda (λ).

The proper integration of hydro and thermal generation for minimum overall cost is quite complex and can be solved optimally only by a digital computer. Even with a computer, the number of calculations used to determine the most economic operation can be so great that considerable computer time is required to obtain a correct solution to the problem.

TRANSMISSION LOSSES

The preceding discussion has centered on determining the loads to be placed on thermal and hydro units in order to obtain equal incremental fuel cost for minimum overall cost of generation. The problem is only partially solved, however, until transmission losses are considered.

It was mentioned previously that if transmission losses could be evaluated, their effect could be used as a multiplier on fuel cost (or water value for hydro) to compensate for the energy lost in transmission and to arrive at a true economic loading of the system.

In the sections on energy transfer and var flows, it was pointed out that all transmission lines have resistance, determined by the conductor material, conductor size, and length of the line. It was also pointed out that the transmission loss in watts was the product of the line current squared times the resistance of the line ($I^2 R$).

In the simplest possible system, a generating unit connected by a single transmission line to a load, the determination of transmission loss is quite simple. Figure 4-7 illustrates this case.

The generator must, of course, produce enough energy to supply the load plus the transmission losses—in the above case, the load plus 100 kW. The power required to supply the losses will move the generation to a higher point on the incremental cost curve, resulting in an increase in the cost of each kilowatthour of energy.

When two or more generating units are connected to a load via separate transmission lines, the correct allocation of load between the units will result when the incremental costs, including the costs of supplying the energy for transmission losses, are equal.

Here again the problem rapidly compounds in complexity as the number of generators, lines, loads, and tie points is increased. Manual methods of calculating loss factors become impractical, and it is necessary to resort to analog or digital computing devices to determine the effects of transmission losses on a power system.

No effort will be made here to develop the mathematical solution of the transmission-loss problem. For the purposes of this discussion, it should suffice to state that a coordination equation has been developed to determine what is called the "penalty factor." Penalty factor is equal

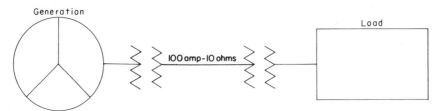

figure 4-7   Simple transmission system of a single generator and load connected by a line carrying 100 A through 10 Ω. Loss is equal to $(100)^2 \times 10 = 100,000$ W or 100 kW.

to $1/(1 -$ loss factor), and it can be seen that as the loss factor increases, the penalty factor will increase.

In order to determine penalty factors, it is necessary to develop a mathematical model of the system. After this has been done, an analog penalty-factor computer or a digital computer can be used to determine penalty factors for any load condition for each generating station or tie-line source to the system load center. When penalty-factor calculations are made "offline," they are manually applied to incremental slide-rule slides for each unit or to penalty-factor setters on analog dispatch-control units. By this means the incremental-cost curves are adjusted upward or downward as required by the penalty factor so that the generating units are loaded on a strictly competitive basis for minimum cost, including transmission losses. These methods have become relatively obsolete because of the wide acceptance and application of digital computers for power system control.

When digital computers are used for system control, penalty-factor calculations are made at frequent time intervals, and generation-control impulses are produced, including current penalty factors, so that the system generation is consistently maintained with the most economic allocation between generating units.

It has been shown previously that minimum fuel input occurs when generating units are operated at equal incremental costs. To demonstrate the effect of transmission penalty factors on load division between gen-

TABLE 4-3

	Unit A			Unit B		
Load, MW	Incr. fuel cost, dollars/MWh	Fuel cost, dollars/h	Load, MW	Incr. fuel cost, dollars/MWh	Fuel cost, dollars/h	Total fuel cost, dollars/h
0	15.50	730	100	60.00	4410	5140
10	17.50	830	90	52.50	3800	4630
20	21.00	1050	80	47.50	3360	4410
30	26.30	1270	70	42.50	2830	4100
40	30.00	1575	60	36.50	2520	4095
47	33.00	1780	53	33.00	2220	4000
50	35.00	1870	50	31.50	2180	4050
60	43.00	2250	40	26.00	1890	4140
70	52.20	3320	30	21.00	1650	4970
80	66.00	4100	20	21.00	1470	5570
90	96.50	5200	10	21.00	1230	6430
100	—	—	0	21.00	1050	—

erating units, an example will be worked out using the two machines previously considered, but with a penalty factor of 1.2 applied to unit B and a penalty factor of 1.0 applied to unit A.

Under these conditions the values shown on the input-output and incremental-cost curves of unit B will be multiplied by 1.2 and replotted. This has been done, and the curves for unit B operating with the assumed penalty factor are shown as the dashed curves on Figs. 4-5 and 4-6. The effect is to raise both the input-output and incremental-cost curves. If the penalty factor had been less than 1, it would indicate that system losses would be reduced by adding load to unit B, and the curves would move downward.

The comparative tabulation under the new operating conditions is shown in Table 4-3. This table shows that the minimum fuel cost occurs with 47 MW on unit A and 53 MW on unit B, with an equal incremental fuel cost of \$33/MWh.

ECONOMIC INTERCHANGE OF POWER

Another problem that is encountered by a system operator is to determine when it is economical to buy power from or sell power to other systems. Whenever power is purchased and received into a system, the power that must be produced to carry the system load is reduced by the amount of power received from the other system. Conversely, whenever power is sold, power production must equal the system load plus the amount of the sale.

The preceding discussion has demonstrated that when the power output of generating units is increased, the unit incremental cost and also the system incremental cost (lambda) increase. Conversely, when power is received from another system, as unit loading is decreased the system lambda decreases.

When power is sold, the additional (incremental) production cost must be determined in order to be able to quote a price to the prospective purchaser of the power.

When power is purchased, production costs will be reduced, and this saving has a value that must be determined. The value of saving in a purchase transaction is called the "decremental value."

Definitions of these two terms are as follows:

1. Incremental cost is the additional cost incurred to generate an added amount of power.
2. Decremental value is the cost saved by not generating an amount of power.

The units usually used are cents per kilowatthour or dollars per mega-watthour.

The method used to determine the incremental cost of a sale trans-action is to take the average of the existing system incremental cost and the new incremental cost and to quote this average figure to the pro-spective purchaser. As an example, assume that the existing cost is $0.03/kWh. If a sale of 100 MW is contemplated, the cost with the new system load condition would be $0.035/kWh. The average incremental cost would then be ($0.03 + 0.035)/2 = $0.0325/kWh.

Exactly the reverse process is used when a power purchase is consid-ered. Assume that the existing cost is $0.03 kWh, and it is desired to purchase 100 MW of power. This amount of received power would reduce the system cost to $0.025/kWh. The decremental value (average saving) would be ($0.03 + 0.025)/2 = 0.0275/kWh.

In considering transactions involving the purchase or sale of power, as in determining how generating units should be loaded for maximum economy, the effect of transmission losses must be considered. As has been pointed out, to determine properly how generating units should be loaded, the unit incremental cost is multiplied by the penalty factor to calculate the worth of the power at the system load center.

When power is being received from another system via a tie line, it is handled exactly as though it were coming from a generating unit at the tie point. The price at the tie point is multiplied by the penalty factor to determine the worth of the purchased power at the load center as compared with that from generating units in the system.

When a power sale is being evaluated, the reverse is true. In this case power is being transmitted from the load center to the tie point with the purchasing system, and it is desired to determine the worth of the power at the tie point. In order to make this determination, the value of the power (system incremental cost) at the load center is divided by the penalty factor.

Examples of both situations will be shown. First, assume a system incremental cost of $0.03/kWh at the load center. A purchase of 100 MW is being considered at a quoted price of $0.026/kWh. The penalty factor from the tie point to the load center has been determined to be 1.15.

To evaluate properly the economics of the proposed purchase, it will be necessary to determine both the cost of the purchased power at the load center and the decremental value of the purchase to the system. The cost at the load center would be $0.026 × 1.15 = $0.0299/kWh. If the system generation is reduced by 100 MW due to the purchase and the system cost is reduced to $0.027, the decremental value would be

($0.03 + 0.027)/2 = $0.0285/kWh. In this case there would be no saving in purchasing the power because the cost of the purchased power is greater than the decremental value of the saving.

Another situation might be developed in which a system with an incremental cost of $0.03/kWh at existing load was asked to supply 100 MW to another system with an incremental cost of $0.042/kWh at its existing load. Assume that the selling system's incremental cost went to $0.035/kWh with the additional load, and that the penalty factor to the tie point is 1.02 at 100-MW delivery. The quoted price would be

$$\frac{\text{Original cost + new cost}}{2} \times \frac{1}{\text{penalty factor}}$$

or $\quad \dfrac{0.03 + 0.035}{2} \times \dfrac{1}{1.02} = 0.0318/\text{kWh}$

The purchasing system would determine its decremental value as follows. Assume that if its generation is reduced by 100 MW, its system cost will be reduced to $0.038/kWh and that the penalty factor from the tie point to load center will be 1.05. The decremental value will be ($0.042 + 0.038)/2 × 1.05 = $0.042/kWh. The difference between the buyer's decremental value and the seller's incremental cost will be $0.042 − 0.0318 = 0.0102/kWh.

In purchase and sale transactions of the type discussed above, it is customary to split the savings between the buying and selling systems. In other words, the average of the sum of the buyer's decremental value and the seller's incremental cost would be determined as in the case just outlined.

$$\begin{aligned} \text{Buyer's decremental value} &= \$0.042/\text{kWh} \\ \text{Seller's incremental cost} &= \$0.0318/\text{kWh} \\ \text{Average} = \frac{0.042 + 0.0318}{2} &= \$0.0369/\text{kWh} \end{aligned}$$

The purchasing system would pay $0.0369/kWh and would save the difference between what it would have cost to generate the power and the cost of the purchased power. In this case $0.042 − 0.0369 = $0.0051/kWh, which represents, at 100-MW delivery, a saving of $510 per hour. The seller would benefit by the same amount.

Obviously when there is a significant difference between costs on systems, it is mutually desirable to enter into transactions of the type just discussed. These are usually termed "economy energy" transactions. Most contractual arrangements between systems for economy energy have a minimum difference, such as $0.005/kWh, before such transactions are permitted. This is to protect against inaccuracies in estimating load that may fluctuate during a transaction period, which is usually for a fixed period such as an hour. Furthermore, the determination of incremental costs and decremental values may not be precise, and unless a significant difference exists, losses rather than savings may result.

In some cases, instead of averaging the seller's incremental cost and the buyer's decremental value, power sales are made by multiplying the seller's incremental cost by a fixed percentage, such as 15 percent. For example, a sale might be effected at seller's incremental cost of \$0.03/ kWh × 115 percent, expressed as \$0.03 × 1.15 = 0.00345/kWh. This method of calculating energy cost is used when the buyer may not know the decremental value. Contracts sometimes permit transactions by either method.

SUMMARY

This section might be summarized by restating that the objective in power system operation is to produce and transmit power to meet the system load at minimum production cost with proper consideration of the effect on system security. This objective is achieved when all generation is operated at equal incremental cost with consideration for transmission losses.

Hydro plants can be integrated into the operation by putting a value on the water used through the hydro plants and then loading them incrementally in competition with the thermal plants.

Material savings in fuel cost can be achieved by careful adherence to economic (incremental) loading of generating units.

Purchase and sale of energy can also be used to minimize power production costs. In considering such transactions, the cost of the purchased power must be compared against the saving that will result by not producing the amount of power involved in the purchase.

PROBLEMS

1. When alternative sources of energy are available to a power system, they should be used in such a way that:
    (a) Thermal generation is held at a minimum.
    (b) The most efficient plants are always loaded to their maximum.
    (c) Overall production cost is minimized.
2. In a thermal-electric generating plant, overall efficiency is improved when:
    (a) Boiler pressure is increased.
    (b) The differences between initial pressure and temperature and exhaust pressure and temperature are held at a maximum.
    (c) Load on the units is increased.

3. When load on a thermal unit is increased, fuel input
   (a) Increases
   (b) Does not change
   (c) Decreases
4. Incremental-heat-rate curves, for thermal generating units, are used to determine the
   (a) Fuel cost in dollars per hour
   (b) Values to which the units should be loaded to result in minimum fuel costs
   (c) Cost per unit of electrical output
5. When generating units are loaded to equal incremental costs:
   (a) Minimum fuel costs result.
   (b) Fuel costs are at a maximum.
   (c) Fuel costs are not affected.
6. One advantage of computer control of generating units is that:
   (a) Var output of the units is minimized.
   (b) All units under the control of the computer will be loaded to the same load.
   (c) Loading of the units will be frequently adjusted to maintain them at equal incremental fuel costs.
7. If the fuel cost of one unit, operating in parallel with other units, is increased and it is desired to maintain fuel cost, the load on the unit will be
   (a) Increased
   (b) Held constant
   (c) Decreased
8. In a power system using both hydro and thermal generation, the proportion of hydro generation can be increased by
   (a) Increasing the price (gamma) of the water
   (b) Reducing the price of the water
   (c) Increasing the field currents of the hydro generators
9. In economic operation of a power system, the effect of increased penalty factor between a generating plant and system load center is to
   (a) Decrease load on the generating plant
   (b) Increase load on the plant
   (c) Hold the plant load constant
10. In a power system in which generating plants are remote from the load center, minimum fuel costs occur when:
    (a) Equal incremental costs are maintained at the generating station buses.
    (b) Equal incremental costs are referred to system load center.
    (c) All units are operated at the same load.
11. Power system A is considering the purchase of 100 MW from system B. The system A cost without the purchase is \$0.04/kWh and with the purchase is \$0.035/kWh. The system B cost is \$0.03/kWh without the sale and \$0.0350/kWh with the sale. On a split-saving basis the saving to system A by purchasing the 100 MW from system B would be
    (a) \$0.035/kWh
    (b) \$0.04/kWh
    (c) \$0.0031/kWh
    (d) \$0.0025/kWh

5

Power System Control

INTRODUCTION

Control of power systems is one of the major responsibilities of power system operators. System voltage levels, frequency, tie-line flows, line currents, and equipment loading must be kept within limits determined to be safe in order to provide satisfactory service to the power system customers.

Voltage levels, line currents, and equipment loading may vary from location to location within a system, and control is on a relatively local basis. For example, generator voltage is determined by the field current of each particular generating unit; however, as has been pointed out previously, if generator voltages are not coordinated, excess var flows will result.

Similarly, loading on individual generating units is determined by the throttle control on thermal units or the gate controls on hydro units. Each machine will respond individually to the energy input to its prime mover.

Transmission-line loadings are affected by power input from generating units and their loadings, the connected loads, parallel paths for power to flow on other lines, and their relative impedances.

It is of course essential for system operators to monitor voltages and line loads continuously at various locations and to take necessary action by raising or lowering voltage or altering generation loading to keep system lines and equipment operating within rated limits.

There is considerable written information on generator voltage regulators, transformer tap-changing devices, and governing systems for control of loading of generating equipment. Consequently, these devices will not be discussed in detail in this section.

POWER SYSTEM CONTROL FACTORS

In considering the more general problem of power system control, the common quantities that affect the entire system can be identified as system frequency and, on interconnected systems, the tie-line flows. Another quantity that would be a valuable guide to system performance would be the angular displacement within the system, as discussed in the section "Transfer of Energy." However, equipment to measure angular displacement accurately has not been developed.

As has been pointed out previously, almost all power systems make use of alternating current. Except for minor momentary excursions of frequency when a generator increases or decreases load with its attendant power-angle changes, the frequency is the same at all points in the system. Consequently, frequency is a basic quantity that can be measured and applied to the control of generating units. Furthermore, since almost all generating units are of the synchronous type, they are locked together at synchronous electrical speed.

When system frequency increases or decreases, the connected generating units will increase or decrease in speed by the same amount electrically. This means that if frequency increases from 60 to 60.1 Hz, all interconnected generators will increase in speed to operate at 60.1 Hz. Of course, the physical speed change will be determined by the number of poles in the machine according to the following formula:

$$\text{rpm} = \frac{120f}{\text{NP}}$$

where
$$\text{rpm} = \text{revolutions per minute}$$
$$f = \text{frequency in hertz}$$
$$\text{NP} = \text{number of poles}$$

For example, at 60 Hz a two-pole machine would operate at (120 × 60)/2 = 3600 rpm, and at 60.1 Hz it would operate at (120 × 60.1)/2 = 3606 rpm.

This would be typical of a steam-turbine-driven alternator. Hydro units operate at much slower speeds. For example, an 18-pole machine at 60 Hz would operate at 400 rpm, and at 60.1 Hz the speed would increase to (120 × 60.1)/18 = 400.67 rpm. It should be emphasized that although the two machines in the above examples operate at radically different physical speeds, the electrical speeds are identical.

FREQUENCY CONTROL

Because system frequency is common to all parts of the system and is easily measured, it was the first quantity applied to system control. The governors on generating units make use of rotating flyballs. These actuate a hydraulic system to open or close the throttle valves of the prime movers of the machines. This action increases or decreases energy input (fuel in a thermal plant or water in a hydro plant) to maintain speed (frequency) at the desired value. More recently electronic governors have been applied that sense frequency and actuate hydraulic devices to control gate or throttle position without the use of flyballs.

In order to operate machines in parallel with stability, it is necessary that the governors have drooping characteristics. That is, as load increases, speed decreases. Governor droops are expressed in percentage of speed change from no load to full load. For example, with a 5 percent droop (a common setting), the no-load speed would be 105 percent of the full-load speed. This is shown graphically in Fig. 5-1.

Since generators operated in parallel cannot be separated to adjust the governor, each time a load change is made, the governor droop characteristic is adjusted during a series of tests and is then left fixed. In operation the speed motor on the governor control system will move the speed controls up or down to correspond to desired load settings, as indicated by curves A and B of Fig. 5-1.

If governors had zero droop, or if they were adjusted so that the speed characteristic increased with load, operation would be unstable. This situation would be similar to overcompounding of dc generators operating in parallel. If one machine has a lower governor droop setting than the others, when two or more generating units are operated in parallel on an ac system, on a frequency drop the machine with the lower droop characteristic will pick up proportionally more load.

Because governors are a combination of hydraulic and mechanical components, an appreciable change in system speed is required before

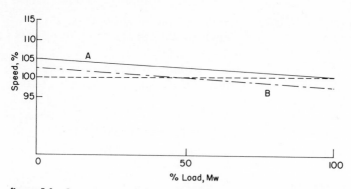

figure 5-1 Governor speed load characteristic. On curve *A* the governor speed motor is adjusted so that at no-load and separated the machine will run at 105 percent speed and at full load at 100 percent (synchronous on system) speed. Curve *B* shows the condition for synchronous speed at 50 percent load. In this case, the no-load, separated speed would be 102.5 percent.

the governor can sense it and take corrective action. Consequently the correction is delayed by a discrete time interval from the time the speed (frequency) change occurred. As a result, machines or systems controlled only by governors have a "dead band" of the order of ± 0.02 cycle. In other words, on a 60-Hz system with governor control only, normal speed will vary between approximately 59.98 and 60.02 Hz.

During system disturbances due to line troubles or load changes, frequency deviations in systems using only governor action will vary depending on the size of the system and the magnitude of the load change or generation change.

For many years governor speed control was the only available means of controlling system frequency. When a load change occurs on a unit operating alone or on a system with governor control only, the speed will stabilize at that indicated for the new load condition. For example, if the unit with the speed load characteristic shown in Fig. 5-1 was operating at 60 Hz at 100 percent load, and load was decreased to 50 percent, it would operate at excess speed, as indicated on curve *A*, until the governor speed control was readjusted so that it could operate on curve *B*.

Automatic generation control systems transmit control pulses to the governor motor operators on the control valves, and cause load to be increased or decreased, as required, to restore the correct frequency. In such systems the governing burden is divided among several generating units, minimizing the amount of gate movement on individual machines. Such electronic AGC systems can control frequency to much closer limits than can mechanical governors.

INTERCONNECTED OPERATION

As pointed out in the section "Economic Operation of Power Systems," significant savings can result from interchanging energy between systems where there is an appreciable difference in generation costs of the systems considering such transactions. For this and other reasons that will be discussed later, there has been a great deal of interconnection between power systems, so that large interconnected power pools have been developed.

Although there are several advantages to power system interconnection, more stringent requirements on load and frequency control must be imposed if pool operation is to be successful. Without precise control of generation and frequency, undesired tie-line flows will result. In addition, the effects of troubles on one of an interconnected group of systems will be seen on the other systems.

Another factor requiring improved power system control has been the development of industrial processes for which precise control of power system frequency has become important. As a result load-frequency control, centralized in the system dispatcher's offices, is now almost universal.

When systems are interconnected, tie-line flows as well as frequency must be controlled. The normally accepted philosophy for interconnected operation is that

1. Each system (control area) should provide enough capacity to carry its expected load at the desired frequency with provision for adequate reserve and regulating margin.

2. Each system (control area) should operate in such a way that it will not impose a regulating burden (changes in generation resulting from load changes in an adjacent system) on intermediate systems.

3. Each system (control area) should continuously balance its generation against its load so that its net tie-line loading agrees with its scheduled net interchange plus or minus its frequency-bias obligation.

MODES OF TIE-LINE OPERATION

In general there are three modes in which interconnected operation can be effected. These are:

1. Flat frequency
2. Flat tie line
3. Tie line with frequency bias

Isolated systems inherently operate in the flat frequency mode because frequency is the only quantity that is affected when load changes. When systems are interconnected, they must operate at the same electrical speed, and speed changes in one system appear in all the interconnected systems. When one system of an interconnected group senses and responds to frequency changes only, it can exert no control over flow on interconnecting tie lines. This is the condition for flat frequency operation.

When a system responds to tie-line-flow changes only and does not respond to frequency changes, it will maintain the desired tie-line flow but will not respond to changes in frequency. This is the mode of operation known as "flat tie line."

Neither of the above modes of operation satisfies all of the three conditions previously listed as desirable for successful interconnected operation. As a result, in North America it is almost universal for interconnected systems to operate in the tie-line-bias mode. When operating with tie-line bias, systems will respond to both frequency changes and tie-line-flow changes and will help to maintain desired frequency and tie-line schedules.

In order to make it possible for a system to respond to frequency and tie-line changes, it is necessary to provide equipment that will develop error signals proportional to the deviations of these quantities from the desired values. In Fig. 5-2 a method of developing such signals for three interconnected systems is shown.

In the diagram it can be seen that tie-line flows and system frequency are made available to the controllers of each system. Each of the controllers compares desired total tie-line flow and desired frequency with the actual quantities and develops error signals. The error signals are used to develop control signals to prime-mover governors, which restore

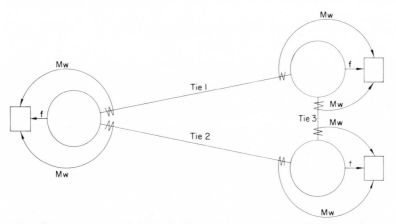

figure 5-2 Diagram of three interconnected systems with tie-line flows telemetered, and a frequency signal applied to the controllers of each system.

tie-line flows to schedule and frequency to normal; that is, they reduce the errors to zero.

Any flow on tie lines above or below scheduled amounts is called tie-line error. That is, the system (or area) is not producing exactly the amount of power required to satisfy its own load plus the amount it is scheduled to deliver or minus the amount it is scheduled to receive.

TIE-LINE BIAS

If frequency deviates from desired frequency (60 Hz), the difference between desired frequency and actual frequency is the frequency error. A loss of generation or fault will cause the frequency of a system to sag, the amount being dependent on the size of the system (rolling inertia, connected load, etc.). The frequency error signal should be adjusted to provide governor control to correct for the frequency swing. This is called frequency bias and is usually designated to megawatts per one-tenth cycle. The bias is a negative quantity because the slope of the governor characteristic curve is negative; that is, the speed of a generator on governor control decreases as load increases, as shown in Fig. 5-1. Figure 5-3 shows the variation of bias correction versus frequency for 50 and 100 MW/0.1 Hz.

The sum of tie-line and frequency errors can be expressed mathematically as "area requirement" or "area control error" as follows:

$$ACE = (T_1 - T_o) - 10B \; (F_1 - F_o)$$

where ACE = area control error (area requirement)

T_o = scheduled net interchange, which normally has a positive sign with power flow out (net tie-line flow, megawatts at normal frequency)

T_1 = actual net interchange (tie-line flow, megawatts)

F_o = desired frequency, hertz

F_1 = actual frequency, hertz

B = area bias, megawatts per one-tenth cycle (which has a negative sign due to the negative slope of the bias characteristic curve)

As an example, assume that scheduled net interchange (net tie-line flow) is 200 MW flowing into the system. Actual net interchange is 150 MW flowing into the system. Desired frequency is 60 Hz, and actual frequency is 60.05 Hz. The bias setting is 50 MW per 1/10 Hz. The area control error would be

$$ACE = [-150 - (-200)] - 10 \times (-50)(60.05 - 60)$$
$$= 50 + 500 \times 0.05 = 75 \text{ MW} \quad \text{(overgeneration)}$$

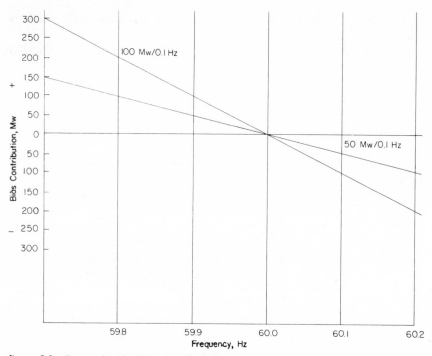

figure 5-3 Curves showing bias contribution versus frequency for 50 MW and 100 MW per 0.1 Hz frequency change.

When a system is operating with tie-line bias control, it will respond to both tie-line-flow errors and frequency errors and assist in achieving the objective of having each control area match its generation with its load. It will also assist in restoring frequency to an interconnected area when an interconnected system suffers loss of generation or transmission.

The frequency-bias setting must be reviewed from time to time to ensure that it is correct. For example, the addition of a large generating unit on a system would increase the inertia of the system and necessitate an increase in bias setting.

When the frequency bias is too low, the system will not respond adequately to take its fair share of total interconnected system control during trouble conditions, resulting in a control burden on other systems. If bias is set too high, overcontrol will result, also putting an excessive control burden on adjacent systems. When the bias is properly set to the natural load-frequency governing response of a system, the system will respond in such a way that area control error will be held to a minimum without imposing an undesired regulating burden on adjacent systems.

ACCUMULATED FREQUENCY ERROR

The frequency-error signal used in system control is developed by comparing system frequency against a frequency standard whose frequency output is not affected by power system operation. Sources of standard frequency are quartz crystals (such as are used to control the frequency of radio transmitting stations), precisely controlled tuning forks, and radio signals from standard-frequency radio stations (WWV) of the Bureau of Standards.

By accumulating the instantaneous frequency deviations, it is also possible to determine the accumulated time error, which is usually limited to not more than 2 s fast or slow. When accumulated time error reaches the limit, or at other convenient times, by agreement, the interconnected systems all offset frequency by a predetermined amount (usually 0.02 Hz), in such a direction that time error will be reduced to zero. By this means the frequency of the interconnected systems is restored to normal. This subject will be discussed further in connection with the control and correction of inadvertent energy accumulations.

SUMMARY

Modern system-control equipment provides control impulses as described above for load-frequency control and also for economic allocation of generation, which was discussed in the section "Economic Operation of Power Systems." Almost all modern automatic generation control (AGC) systems make use of digital computers. Analog automatic dispatch systems have become relatively obsolete.

Digital systems have replaced analog systems: (1) because of their capabilities for handling computational work simultaneously with system control, and (2) because of the relative ease with which their control functions can be expanded or altered, frequently requiring only programming changes.

Frequently AGC is combined into supervisory control and data acquisition systems, which are discussed in more detail in Chap. 8.

Before leaving this subject, it might be well to stress again that the objectives of system control are as follows:

1. Each system should provide enough capacity to carry its expected load at the desired frequency with provision for adequate reserve and regulating margin.

2. Each system should operate in such a way that it will not impose a regulating burden on interconnected systems.

3. Each system should continuously balance its generation against its load so that net tie-line loading agrees with its scheduled net interchange plus or minus its frequency-bias obligation.

PROBLEMS

1. Governors for controlling the speed of electric generating units normally provide
   (a) A flat speed load characteristic
   (b) An increase of speed with increasing load
   (c) A decrease of speed with increasing load
2. When two identical ac generating units are operated in parallel on governor control, and one machine has a 5 percent governor droop and the other a 10 percent droop, the machine with the greater governor droop will
   (a) Tend to take the greater portion of load changes
   (b) Share load equally with the other machine
   (c) Tend to take the lesser portion of load changes
3. On load-frequency-control installations, error signals are developed proportional to the frequency error. If frequency declines, the error signal will act to
   (a) Increase prime-mover input to the generators
   (b) Reduce prime-mover input to the generators
   (c) Increase generator voltages
4. When two or more systems operate on an interconnected basis, each system
   (a) Can depend on the other systems for its reserve requirements
   (b) Should provide for its own reserve capacity requirements
   (c) Should operate in a "flat frequency" mode
5. When interconnected power systems operate with tie-line bias, they will respond to
   (a) Frequency changes only
   (b) Both frequency and tie-line load changes
   (c) Tie-line load changes only
6. Power system A is operating in parallel with other power systems and has a frequency bias of 50 MW/0.1 Hz. For a loss of generation in one of the interconnected systems which results in a frequency drop of 0.03 Hz, power system A should
   (a) Provide 30 MW of support
   (b) Not be affected
   (c) Provide 15 MW of support
7. Area control error is given by the formula

$$\text{ACE} = (T_1 - T_o) - 10B \ (F_1 - F_o)$$

where
   ACE = area control error
   T_o = scheduled net interchange
   T_1 = actual net interchange
   F_o = desired frequency, hertz
   F_1 = actual frequency, hertz
   B = area bias, megawatts per one-tenth hertz

If scheduled net interchange T_o = 200 MW into the system, actual net interchange T_1 = 180 MW into the system, F_o = 60 Hz, F_1 = 59.9 Hz, and B = 150 MW/0.1 Hz, the area control error is
   (a) + 130 MW
   (b) Zero
   (c) + 170 MW
   (d) − 130 MW

8. If frequency bias is set too low, an interconnected power system will respond to troubles in adjacent systems with

(a) More than its share of bias response

(b) Correct bias response

(c) Less than its share of bias response

9. If a power system operates at an average frequency of less than 60 Hz:

(a) No time error will accumulate.

(b) There will be a slow accumulated time error.

(c) There will be a fast accumulated time error.

10. A power system operator who observes an accumulated time error should correct it by

(a) Increasing generation on the system

(b) Reducing generation on the system

(c) Coordinating time-error correction with other interconnected systems

6

Energy Accounting
In Interconnected
Operations

INTRODUCTION

Measurement of electric energy supplied to customers or purchased from and delivered to interconnected power systems is of paramount importance in power system operation. Energy delivered to customers must be accurately measured so that each customer can be billed for the amount of energy actually delivered. Energy transferred between systems must be properly measured and accounted for to ensure that agreed-upon schedules are being met.

Before the development of integrating watthour meters, flat rates, determined by the number of lamps or other connected load, were used for billing purposes. There was no means for determining the actual amount of energy used by a customer.

MEASUREMENT OF ENERGY

The revolving-disk watthour meter was developed in the early days of the power industry and is still used to measure energy deliveries or transfers. These devices have not changed basically, although there have been many improvements in their design and construction.

As ordinarily constructed, they have potential (voltage) elements with 115- or 230-V windings and current elements determined by the connected load for usual domestic applications.

To measure energy at the high voltages and currents used in power systems or with large electrical machinery, it is necessary to reduce the voltages and currents to values that the watthour meters can handle. For this purpose potential and current transformers with specific ratios of turns between the primaries and secondaries are used.

Potential transformers for power system applications will typically have winding ratios such as 300:1 for 60-kV applications, 600:1 for 115-kV, 1200:1 for 230-kV, etc. Potential transformers are usually connected line to ground so that the line-to-line voltage is the indicated meter voltage times the transformer ratio times 1.73 ($\sqrt{3}$). For example, on a 115-kV system with a 600:1 potential transformer and 120 V indicated on the meter, the actual line-to-line voltage would be 120 × 600 × 1.73, or 124,500 V (124.56 kV).

Current transformers have their primary windings installed in series in the lines in which currents are being measured. The primary windings consist of very few turns of a conductor of adequate size to carry maximum current in the line. The secondary winding consists of many turns of a small conductor with a turns-winding ratio, expressed in primary amperes to secondary amperes, such as 100:5, 600:5, 1000:5 A, etc. Actual turns ratios of the above transformers would be 20:1, 120:1, and 200:1.

All transformers are actually voltage-operated devices. In current transformers, with rated line current flowing in the primary winding, there will be a voltage drop across the winding sufficient to induce a voltage in the secondary winding adequate to cause rated secondary current to flow. It might also be in order to point out that current-transformer secondary windings should always have a closed circuit across the secondary terminals. Due to the high ratio of secondary to primary turns, extremely high voltages can result in current-transformer secondary windings if they are open-circuited.

There are various connection arrangements for metering three-phase power. Usually either the "three-wattmeter" or "two-wattmeter" method is used. These methods are well described in electrical textbooks and will not be discussed here. This discussion is primarily concerned with measuring energy at voltages and currents used in power systems which require the use of instrument transformers (current and potential), as described above.

It should be obvious that when current and potential transformers are used, the metered indication will have to be multiplied by the ratios of both sets of transformers to determine the energy actually delivered.

As an example, assume the metering installation shown in Fig. 6-1. With the instrument transformers shown, voltage and current readings would be multiplied by

$$\frac{(400/5) \times 600 \times 3}{1000} = \frac{80 \times 600 \times 3}{1000} = 144$$

to determine the actual power in kilowatts in the circuit. For example, if the load current reading was 4 A and the voltage reading was 120 V, the power in the circuit, assuming unity power factor, would be 69,120 kW.

Ammeters used in connection with current transformers are usually calibrated to read actual line current, taking into consideration the current transformer ratios. Voltmeters frequently read potential transformer secondary voltage. In order to obtain actual line voltage it is necessary to multiply the indicated reading by the potential transformer ratio. However, the meters may be calibrated to read actual line voltage.

Indicating watt and var meters are frequently installed on transmission-line terminals so that operators can see real and reactive power flows and the effect on reactive flow of generator field or other voltage-control-device changes.

In measuring energy, time becomes a factor. The product of current, voltage, and power factor gives power in a circuit only at the instant at which the readings were taken. However, if current and voltage remain constant for a period of time, such as 1 h, the energy used or transported

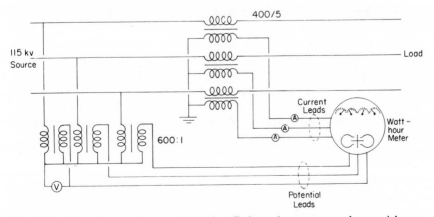

figure 6-1 Typical three-phase metering installation using current and potential transformers.

will be the indicated watts times 1 h, or watthours. Mathematically expressed, energy is watts × time (hours) = watthours.

Since a watthour is a relatively small unit, it is usual to express energy in kilowatthours or megawatthours. One kilowatthour = 1000 watthours and 1 megawatthour = 1,000,000 watthours.

Measurement of energy is accomplished by the use of integrating watthour meters. Essentially such meters consist of an electric motor whose speed of rotation is proportional to the amount of power flowing in the circuit. The shaft of the motor rotor (disk) is connected through a gear train to indicating dials. The watthour meter adds up all the instantaneous powers in a time period to indicate the energy (power × time) used by a load or transported in a circuit in the time period.

As previously mentioned, watthour meters usually have current and potential coils of relatively low rating, and in power system applications are used with current and potential transformers. In addition to the multiplying factor resulting from the instrument transformers, the gear train from the meter disk to the indicating dials also introduces a constant multiplying factor. By proper selection of the gear train, the total multiplying factor can be made to equal some convenient constant, such as 1000 or 10,000.

The energy measured by an integrating watthour meter is always obtained by taking the difference between a present reading and a previous reading and multiplying by the meter constant.

Energy supplied to customer loads is usually determined directly by integrating meter readings. This is also true in determining the energy produced by a generating unit, where daily, weekly, or monthly readings may be taken, depending on the desired time period.

INTERCONNECTED ENERGY ACCOUNTING

A different problem arises in accounting for energy in interconnected operation between power systems. The watthour meters are located at tie points, which usually are remote from the system operations centers. It is possible, of course, to telemeter the watthour meter readings, but another complication arises when several categories of power are being transported on the same line at one time. For example, surplus and firm power may be scheduled simultaneously, or power for more than one system may be transported over a single tie line at one time. The meters are not capable of distinguishing between the different classes of power and indicating them individually.

As a consequence, it has become customary in interconnected operations to account for and bill for energy on the basis of amounts scheduled for a given time period. The integrating (watthour) meter readings are then used to determine whether or not the schedules have actually been met.

An example might be a schedule in which 100 MW of firm and 50 MW of economy energy are scheduled simultaneously on a line for a period of 1 h. If the schedules are met exactly, the metered difference in the 1 h period should be 150 MWh. If, in fact, the metered difference is 147 MWh, the supplying system has been deficient in its delivery by 3 MWh.

Such a case would be handled as though the receiving system actually received the scheduled energy (150 MWh) and the bill would include this amount at the applicable rates. The 3-MWh deficiency would become a part of an "inadvertent" energy account. This account is maintained to keep track of over or under deliveries due to errors in settings of the tie-line scheduling devices or of the tie-line control equipment.

It should be stressed that limitation of frequency excursions by the control system will minimize the accumulation of inadvertent energy and the time errors that can result.

Even when no power is scheduled across a tie, control equipment is not precise enough to maintain the ties at exactly zero. Also, when there is more than one path through which power can flow, it will divide between systems in proportion to line impedances and power angles,

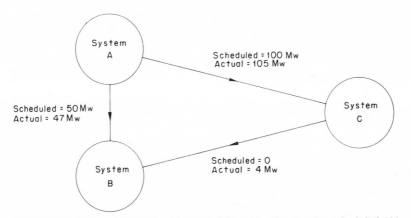

figure 6-2 Diagram of three interconnected systems. System A has scheduled 100 MW to system C and is actually delivering 105 MW. System A has scheduled 50 MW to system B and is actually delivering 47 MW. The system A inadvertent is accumulating at a rate of $+2$ MWh/h. System B has scheduled 50 MW from system A and is actually receiving 47 MW and has scheduled 0 MW from system C and is actually receiving 4 MW. System B inadvertent is accumulating -1 MWh/h. System C has scheduled 100 MW from system A and is actually receiving 105 MW and has scheduled 0 MW to system B and is actually delivering 4 MW. System C inadvertent is accumulating at a rate of -1 MWh/h. However the deviation of the overall power pool is zero. System A ($+2$), system B (-1), and system C (-1).

and some power will flow through a system which has not scheduled receipt or delivery of power.

As a result of tie-line-control inaccuracies or parallel paths, inadvertent energy can accumulate in each interconnected system. How such accumulations can occur is shown in Fig. 6-2.

INADVERTENT ENERGY ACCOUNTING

Inadvertent energy accounts are not permitted to accumulate large amounts. Efforts are made to balance inadvertent energy accumulations from time to time by adjusting the tie-line controller to receive a little more or a little less (depending on whether the inadvertent accumulation is positive or negative) than is scheduled. Such corrections are usually made in connection with time-error correction or by mutual agreement among systems. In the case shown in Fig. 6-2, if system C adjusted its generation so that it actually received 103 MW, and if at the same time the tie to system B went to 5 MW, system C would balance its inadvertent account at a rate of 2 MWh/h.

By convention, the sign of the inadvertent energy accumulations is negative ($-$) when it indicates undergeneration on the part of a system and positive ($+$) when it indicates overgeneration.

Contracts between systems frequently provide that inadvertent energy accumulations shall be balanced during time periods the same as those in which the accumulation occurred. That is, inadvertent energy accumulated during the peak period of the day should be balanced during the peak period. If this procedure were not followed, it would be possible to accumulate negative inadvertent (undergeneration) during the peak period when incremental costs are higher and repay it by overgeneration during "off-peak" hours when incremental generation costs are low. Such a procedure in balancing inadvertent accounts would obviously result in a loss to the system that accumulated a positive (overgeneration) inadvertent during peak periods. Ordinarily inadvertent accounts are kept for two time periods, such as 0800 to 2000 (peak) and 2000 to 0800 (off-peak).

Another problem that is related to inadvertent energy accumulations is accumulated time error. If power system control equipment were capable of providing perfect control, system speed would be maintained exactly and time error would not accumulate. Commercially available control equipment is not capable of providing such exact control, and as a result average frequency will deviate slightly from desired frequency. In a time period of several hours this deviation will result in an accumulated time error proportional to the average frequency error. For example, if frequency averaged 60.01 Hz for an hour, a time error of 0.60 s, fast, would accumulate in this period.

In interconnected operation, of course, all the interconnected systems operate synchronously, and any frequency error will be common to all the systems. If time error only is being corrected, all the systems can simultaneously offset frequency, fast or slow, for a period sufficient to correct the time error without affecting inadvertent energy accumulations.

It is possible to limit or help reduce inadvertent energy accumulation in coordination with time-error correction. For example, if there is a positive (overgeneration) inadvertent accumulation and a fast time error, a system can reduce its generation slightly to permit some inadvertent flow in a direction opposite to the accumulation, and as a result the area average frequency will be slightly reduced, resulting in a gradual reduction in time error.

In the above case, if the system had increased generation slightly, both inadvertent energy and time error would have been increased.

From this discussion it should be apparent that the signs of the inadvertent energy and of the time error must be observed in correcting either of these quantities. The general rule is as follows:

1. If inadvertent energy is positive (overgeneration) and time error is fast, tie-line schedules can be offset in the direction of reducing generation slightly to reduce both the inadvertent energy accumulation and time error.

2. If inadvertent energy is negative (undergeneration) and time error is slow, tie-line schedules can be offset in the direction of increasing generation slightly to reduce both inadvertent accumulation and time error.

Some power systems have automatic time-error correction equipment on either analog or digital-dispatch computer installations which helps to limit both inadvertent energy and time error. As such installations become more common, it should be possible to minimize both these quantities.

SUMMARY

The preceding discussion can be briefly summarized as follows:

1. Energy is the summation of instantaneous power over a period of time and is usually expressed in watthours, kilowatthours, or megawatthours.

2. Energy accounting for interconnected system operation is usually done by considering the amounts scheduled as being actually delivered, and any difference between scheduled and metered actual deliveries as inadvertent energy.

3. A positive (+) accumulation of inadvertent energy indicates overgeneration, and a negative (−) accumulation of inadvertent energy indicates undergeneration.

4. By proper coordination, both time error and inadvertent accumulations can be corrected or minimized.

PROBLEMS

1. A metering installation for a 115-kV tie point between two power systems has 1000:5 current transformers in each phase and 600:1 potential transformers connected line to ground. Assuming unity power factor, the multiplying factor that should be used to determine the power in the circuit in kilowatts is
    (a) 360
    (b) 3600
    (c) 207.8
2. Interchange transactions between power systems are usually accounted for by
    (a) Taking the difference of watthour meter readings at the start and finish of each transaction period
    (b) Taking the average of the delivery at the start and finish of the transaction period
    (c) Considering that the amount scheduled has actually been delivered
3. A power system has scheduled a purchase of 200 MW for 3 h. The difference in watthour meter readings at the start and finish of the delivery indicated that 595 MWh was actually received. As a result of the underdelivery, the power system would
    (a) Ignore the difference
    (b) Reduce its payment to the supplying system
    (c) Include the amount of underdelivery in its inadvertent energy account
4. When one of a group of interconnected power systems generates power in excess of its own area requirement and scheduled tie-line obligations, the interconnected group of systems will be affected by
    (a) Below-normal frequency and inadvertent power flow to the system that is overgenerating
    (b) Above-normal frequency and inadvertent power flow to the system that is overgenerating
    (c) Above-normal frequency and inadvertent power flow from the system that is overgenerating
    (d)Below-normal frequency and inadvertent power flow from the system that is overgenerating
5. Inadvertent energy accumulations between power systems are normally balanced by scheduling compensating deliveries of power
    (a) During off-peak periods
    (b) During periods corresponding to those in which the inadvertent energy was accumulated
    (c) On weekends
6. When positive inadvertent energy has been accumulated by a power system and time error is fast, the system can reduce the inadvertent balance and reduce time error by
    (a) Reducing generation slightly
    (b) Increasing generation slightly
    (c) Increasing the tie-line-bias setting of the load-frequency-control equipment

7. Automatic time-error correction equipment will act to reduce generation
  (a) Whenever time error is slow
  (b) Whenever time error is fast
  (c) When time error is slow and there is a negative inadvertent energy accumulation
  (d) When time error is fast and there is a positive inadvertent energy accumulation

7

Telemetering Methods

INTRODUCTION

The operation of power systems requires that a great deal of information be continuously available to the system operators. Power systems usually cover relatively large geographic areas, and consequently points of generation, interconnection to other systems, and major loads may be separated by many miles. Remote metering is used in order that information on loads being generated or supplied, reactive flows, bus voltages, etc., at key points can be made available for indication or automatic control.

Remote metering or indication is called "telemetering." Various methods of telemetering have been developed and are in common use in utilities and other industries.

Basically, there are two types of telemetering systems—"analog" and "digital." In analog telemetering, a voltage, current, or frequency proportional to the quantity being measured is developed and transmitted on a communication channel to the receiving location, where the received signal is applied to a meter calibrated to indicate the quantity being measured or directly to a control device such as a dispatch computer.

In digital telemetry the quantity being measured is converted to a code in which the sequence of pulses transmitted indicates the quantity. One of the major advantages of digital telemetering is the fact that accuracy of data is not lost in transmitting the data from one location to another. Accuracy of telemetering will be discussed at greater length later.

Modern supervisory control systems are usually computer-based. In such systems, digital messages are used both for control functions and for transmission of data back to the master station. These systems will be discussed in more detail in Chap. 8.

It might be well to examine basic principles of telemetering before discussing various types of telemetry.

In any telemetering system there must be a transducer, a communication channel, and a remote indicating or recording device or output terminals.

VARIABLE-CURRENT TELEMETERING

Perhaps the simplest telemetering system is one in which a potentiometer is used to develop a voltage proportional to the position of the potentiometer slider, which can be moved by a float, cam, or other device controlled by the process or function being measured. An example of a potentiometric telemetering system is shown in Fig. 7-1.

A simple telemetering system such as that shown in Fig. 7-1 has many limitations. The conductor of the line between the transmitting and receiving devices is in series, and the voltage drop of the line must be considered in calibrating the indicating device at the receiving end. Also, line resistance becomes excessive if an attempt is made to transmit indications over significant distances. Temperature also affects line resistance so that the indication on the receiving voltmeter is affected. Systems of this type have very limited application and cannot be practically applied over long distances.

The basic principle of variable-current telemetering is often used for remote measurement of reservoir levels, tap changer position, and other applications where the voltage output from a potentiometer is used to control a variable-frequency oscillator. The varying frequency is transmitted, and when received is again converted to current or voltage by a transducer. It can then be converted to a digital code by means of an analog-to-digital converter, and used as an input to a supervisory control and data acquisition (SCADA) system.

The problems of changing line resistance, voltage drop, and varying temperature are avoided by converting the variable current to a varying frequency. In such systems the potentiometer is a position to voltage transducer, and the voltage is then converted to a frequency.

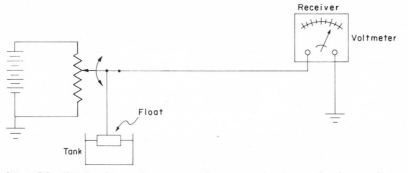

figure 7-1 Simple telemetering system using a potentiometer to develop a voltage proportional to float level. The receiving-end voltmeter can be calibrated to indicate the level in the tank.

PULSE-LENGTH TELEMETERING

A more practical telemetering method is to send a series of pulses on a communications circuit. The transmitting device is constructed in such a way that the length of the pulse is varied in proportion to the quantity being measured. At the receiving end of the system the telemeter receiver converts the variable-length pulses to a voltage, current, or mechanical position proportional to the original quantity being measured. One commonly used method of developing variable-length pulses is to use a motor-driven cam revolving at a constant speed. Contacts are closed for varying lengths of time as determined by the position of an arm, which is moved mechanically or electrically proportional to the quantity being measured. This system is illustrated in Fig. 7-2.

With a telemetering system such as the one described above, the total

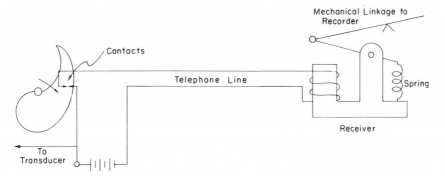

figure 7-2 Pulse-length telemetering system.

period for each sequence of signal and no signal is determined by the speed of rotation of the cam. When the transmitted quantity is zero, a short pulse (signal) will be transmitted, with a period of no signal until the cam rotation has been completed. When a maximum quantity is to be transmitted, the cam closes the contacts for a period of time determined by the position of the movable contact controlled by the transducer, followed by a short period of no signal. The diagram in Fig. 7-3 shows the relationships.

The pulse-length telemetering system is superior to the varying-current system in that communication-line resistance does not appreciably affect the signal. Basically, it and all systems making use of pulses are telegraph systems. Signals of such systems can be transmitted over a pair of wires or superimposed on a carrier for transmission over wires, power lines, or microwave.

One of the disadvantages of the pulse-length system is that it is relatively slow. As commercially available, most systems require several seconds for a complete signal cycle. However, for many purposes this is entirely adequate, and many of these systems are in service.

PULSE-RATE TELEMETERING

Another basic telemetering system is a pulse-rate or variable-frequency system. In such a system the number of pulses or cycles transmitted in a given time period is varied in proportion to the quantity being measured and telemetered. Various devices can be used to cause the pulse rate to change, such as a variable resistor (rheostat) in a resistance-tuned oscillator.

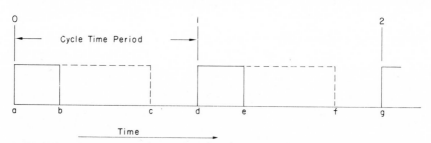

figure 7-3 Diagram of signal pulses for pulse-length telemetering system. Starting at time a with a minimum quantity to be telemetered, a signal is transmitted until time b, and then no signal until time d, when a signal is transmitted until time e, and no signal until time g. Times a–b and d–e are equal. This sequence continues until the quantity to be telemetered changes.

When a maximum quantity is to be telemetered, a signal will be transmitted during the periods a–c and d–f, and no signal during the periods c–d and f–g, etc. Quantities between zero and maximum would have signals transmitted with pulse lengths increasing proportionally up to the maximum.

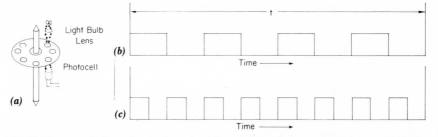

figure 7-4(a)   Method of developing variable pulse rate with a watthour meter element. With a revolving disk pulses are produced at a variable rate as the light beam is interrupted by the disk rotation, which varies in speed proportional to the quantity being measured. (*b*) Four complete pulse cycles during time *t*. (*c*) Eight pulse cycles in the same time period.

One of the methods used to develop a varying pulse rate for telemetering is to make use of what is essentially a watthour meter element with a rotating disk. The disk has notches in its outer periphery or holes perforated in it. A light beam is passed through the notches or holes to a photocell so that when light strikes the photocell a signal is developed. As previously mentioned, a watthour-meter disk will revolve at a rate proportional to the power flowing in the meter circuit. If a constant voltage is supplied to the potential coils, the rate of disk rotation will depend entirely on the amount of current flowing in the current coils, which can be made proportional to the quantity being measured. This method of telemetering is illustrated in Fig. 7-4.

Other devices which are totally electronic or which use magnetic amplifiers have been developed. In these devices there are no moving parts, and a varying frequency, or pulse rate, is developed by applying a voltage, proportional to the quantity being measured, to the device.

PULSE-AMPLITUDE TELEMETERING

Another method of telemetering is to vary the amplitude of the signal. This system is not commonly used because noise on a communication circuit appears as an amplitude change and can give erroneous indications.

DIGITAL TELEMETERING

All the methods of telemetering discussed so far are of the analog type. As previously stated, the second basic type of telemetering uses digital codes to represent the quantity being transmitted by the telemetering system. In digital telemetry the combination of pulses represents numbers or letters corresponding to the measured quantity. In any telegraph system the presence or absence of a signal represents a minimum amount of information and is called a bit.

Different codes have been developed, such as binary and binary-coded decimal, which are two of the most commonly used. In binary coding, numbers are expressed to the base 2, instead of to the base 10 as in the conventional number system. In binary notation a number can be represented as a sequence of ones and zeros, which can be duplicated by periods of signal and no signal in a communication circuit. The development of this form of notation is as follows: Any number to the zero power is 1, so that $2^0 = 1, 2^1 = 2, 2^2 = 4, 2^3 = 8$, etc. In binary notation numbers to 10 would appear as follows:

$$0 = 00000$$
$$1 = 00001 = 2^0$$
$$2 = 00010 = 2^1$$
$$3 = 00011 = 2^1 + 2^0$$
$$4 = 00100 = 2^2$$
$$5 = 00101 = 2^2 + 2^0$$
$$6 = 00110 = 2^2 + 2^1$$
$$7 = 00111 = 2^2 + 2^1 + 2^0$$
$$8 = 01000 = 2^3$$
$$9 = 01001 = 2^3 + 2^0$$
$$10 = 01010 = 2^3 + 2^1$$

The above table could be expanded to represent any number, so that a sequence of signal and no signal corresponding to the ones and zeros can be transmitted. Codes for signals expressing binary numbers would appear as shown in Fig. 7-5.

Usually the digital signals used in a telemetering system are produced electronically by an "analog-to-digital" converter, although it is possible to produce them mechanically through a series of cams or by a group of electrical relays.

Analog-to-digital converters produce the digital signals corresponding to a voltage input which varies proportionally with the quantity being

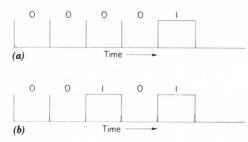

figure 7-5   Representation of binary numbers by signals on a communications circuit. (*a*) Signal representing the binary number 1. (*b*) Signal representing the binary number 5.

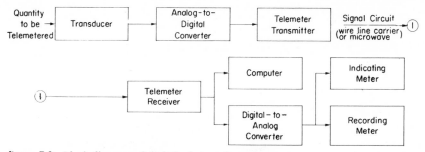

figure 7-6 Block diagram of digital telemetering system.

measured. At the receiving end of the telemetering system, if indication or recording is desired, the signals are converted back to an analog voltage by a digital-to-analog converter.

Where it is desired to make use of such signals in a digital computer, the digital signals can be applied directly to the computer for use in printing a log of values, or for control purposes in the case of process-control computers. A representation of a digital telemetering system is shown in Fig. 7-6.

Because accuracy is not lost in transmission, digital telemetering is gaining wide acceptance. In such systems the overall accuracy is determined by the accuracy of the transducers and the analog-to-digital and digital-to-analog converters used in the system.

BANDWIDTH REQUIREMENTS

In any communications system the width of the frequency band required to transmit information is proportional to the amount of information being transmitted in a given time. For example, a manual telegraph circuit requires a frequency band of only approximately 0 to 100 cycles, a voice telephone channel requires a band of approximately 3000 cycles, and a television channel requires approximately 6,000,000 cycles of bandwidth.

This is also true in telemetering and where high-speed transmission of information is required, where wideband telemetering channels are needed. When digital telemetering is used, the channels are specified in "bits" per second, where a bit is the smallest element of information (signal or lack of signal). The bit rate capability of a circuit is determined as follows:

$$\frac{\text{Words per} \atop \text{minute} \times \text{characters} \atop \text{per word} \times \text{bits per} \atop \text{character}}{60 \text{ s}} = \text{bits per second}$$

A word is considered to be six characters—five letters and a space. As an example, assume 240 words per minute: With six characters per word and 9 bits per character, the bit rate is $(240 \times 6 \times 9)/60 = 216$ bits/s.

CARRIER CURRENT

For transmission of telemeter signals over more than a few miles, it is usually economical to make use of a carrier on wire lines, power lines, or microwave systems. This requires that the signals to be transmitted be superimposed on the carrier. This process is called modulation, and it causes the carrier signal to vary in accordance with the changes in the signal used to modulate the carrier. Conventional broadcast radio uses "amplitude modulation," where the amplitude (power) of the signal is varied in accordance with the information being transmitted.

Noise due to electrical disturbances (static), etc., directly affects the signal amplitude, and in circuits in which noise is high it may not be possible to get useful signals at the receiving end.

FREQUENCY MODULATION AND FREQUENCY-SHIFT TELEMETERING

Frequency modulation causes the carrier frequency to vary in accordance with the applied signal. This method is used for high-quality, noise-free radio. A special form of frequency modulation is frequently used to provide noise-free transmission of telemetering and supervisory control signals. This method is referred to as "frequency shift." As described

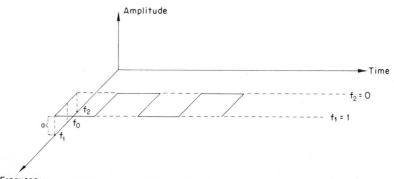

figure 7-7 Diagram illustrating the principle of frequency-shift telemetering. The frequency-shift transmitter operates about, but never at, nominal frequency f_0 and always at constant amplitude a. The presence of a telemeter pulse is indicated when the signal is at f_1, and the absence of a pulse at f_2. Noise would appear as a change in amplitude and has no significance, since it does not contain information.

previously, telemeter signals are usually a series of pulses. With frequency shift, presence of a pulse is indicated by transmission at one frequency and absence of a pulse by transmission at another frequency. Amplitude variations due to noise or other causes have little or no effect on a frequency-shift system. Frequency-shift transmission is illustrated in Fig. 7-7.

As shown this is a non-return-to-zero system. That is, the frequency goes from f_1 to f_2 directly. In some systems a return is made to f_0, so that the signal sequence would be at time $0, f = f_0$. Then one condition would increase frequency to f_1 and then return to f_0, and the other condition would reduce frequency to f_2 and then return to f_0.

SUMMARY

The rapid increase in telemetering, supervisory control, and the application of computers to power system operation will greatly increase the use of telemetering and transmission of control signals on power systems. A basic understanding of the principles involved in transmission of information used for control, indication, and operation of power systems should be of value to everyone involved in power system operation. The preceding discussion is by no means complete, but should serve to make the principles used in telemetering and remote signaling better understood.

PROBLEMS

1. Analog telemetry is accomplished by transmitting signals where the quantity being telemetered has been converted to
    (a) Pulses of varying length or rate or a current proportional to the measured quantity.
    (b) A code for each value of the quantity
    (c) A microwave signal
2. Variable-current telemetering methods are
    (a) Affected by line resistance
    (b) Capable of being used over great distances
    (c) Unaffected by temperature and line length
3. In pulse-length telemetering systems, the length of the pulses
    (a) Is not affected by the value of the quantity being telemetered
    (b) Is proportional to the quantity being telemetered
    (c) Is such that high-speed telemetering is accomplished
4. Pulse-rate telemetering systems
    (a) Are slower than pulse-length telemetering systems
    (b) Are unaffected by noise in the communication channel
    (c) Are developed by causing the number of pulses being transmitted to vary proportionally with the quantity being telemetered
5. In digital telemetry the quantity being telemetered is converted to
    (a) Variable-length pulses
    (b) A code sequence of signal and no signal for each value of the quantity being telemetered
    (c) Pulses of varying amplitude

6. In a binary code the number 7 is represented by a period of signal and no signal as follows:

   (*a*) 01010

   (*b*) 00011

   (*c*) 00111

7. An analog-to-digital converter produces

   (*a*) An analog signal from digital information

   (*b*) Digital signals from analog information

   (*c*) Variable-length pulses proportional to a quantity being measured

8. Digital telemetering is

   (*a*) Less accurate than analog telemetering

   (*b*) Not applicable to control computer installations

   (*c*) Capable of being applied directly to digital control computers

   (*d*) Being supplanted by analog telemetering systems

9. As signaling speed is increased, the bandwidth required

   (*a*) Increases

   (*b*) Is reduced

   (*c*) Is not affected

10. In applications of carrier current for transmission of telemetering signals:

   (*a*) Microwave is always used.

   (*b*) It is necessary to use frequency modulation.

   (*c*) A carrier signal is caused to change in accordance with the information to be transmitted.

11. Frequency modulation is

   (*a*) Less subject to noise than amplitude modulation

   (*b*) More subject to noise than amplitude modulation

   (*c*) Capable of transmitting more information over the same bandwidth

12. When frequency shift is used to transmit telemetering signals:

   (*a*) The frequency-shift signal is varied in amplitude in proportion to the quantity being telemetered.

   (*b*) The frequency is varied in proportion to the quantity being telemetered.

   (*c*) Transmission is alternately at one frequency or another to indicate presence or absence of an information pulse.

8

Supervisory Control and Data Acquisition Systems

INTRODUCTION

Supervisory control systems have been used in power systems for many years. However, developments, particularly in electronics and communications, have made supervisory control and data acquisition (SCADA) systems capable of providing greatly increased performance compared with older systems. Furthermore, increased costs of manual operation and attendance have made it economically attractive for power systems to apply SCADA systems to a much greater extent than was common some years ago.

Almost all modern dispatch and operating centers of power systems are now provided with at least some SCADA system equipment. These types of equipment have proved to be efficient tools for power system and station operators, making it possible for them to maintain relatively complete knowledge of conditions on the power system, or the portions of the system for which they are responsible, as well as to remotely control equipment such as circuit breakers or other devices that are equipped for such control. This chapter briefly describes the functioning of SCADA systems, as well as some of their capabilities and limitations.

CONTROL AND SUPERVISION

The term "supervisory control" is normally applied to remote operation of such devices as motors or circuit breakers, and the signaling back (supervision) to indicate that the desired control action has been accomplished. Simple supervisory control systems have been used since the early days of power utility operations. Early systems consisted of wired control circuits between the control point and the controlled device, along with indicating circuits from the controlled device back to the control point. Control supervision was normally provided by lights, with, for example, a green light to indicate an open device and a red light to indicate that the device was closed.

With such systems, wires between the control point and the controlled device were required for each device being remotely controlled and supervised. In applications in which there were many such devices, the cost and complexity of these systems increased directly with the number of devices to be remotely controlled and supervised. When significant distances were involved, costs were high, and reliability suffered because of induced electrical noise in the control circuits and the possibility of physical failure of the long control circuits.

Some of the limitations of direct-wire circuits between the controlling and controlled equipment on a one-for-one basis were minimized by the use of selective relays, similar to those used in dial telephone systems. By such means it was possible to select the device to be controlled, operate it, and transmit a supervision signal back to the control point over a single control circuit. Such systems became quite complex, however, and were sometimes difficult to maintain. They were also limited in speed of operation, and in maximum data transfer when the number of controlled and supervised devices became large.

The advent of electronic communications methods and digital data transmission provided a means of greatly increasing the capabilities of supervisory control systems. These types of systems could also be made more reliable and were less expensive than the older systems. Sequential scanning of remote stations and of equipment within the stations, supplied with remote supervisory terminal units, made it possible for one master station to control several remote stations and many devices at each remote station. Furthermore, it was possible to telemeter back to the master station the control actions taken by the remote unit, as well as live analog data on current, voltage, power, and many other items needed for the complete supervision of a remote station.

Further advances were made by reducing the amount of data transferred between the remote units and the master station. This is done by a procedure known as exception reporting, where certain data are transferred only when they change or fall outside previously set limits.

These developments made it possible for a central location equipped with a supervisory master station to have almost complete control and information concerning the status of stations under the control of a single master station.

In most systems, the master unit sequentially scans the remote terminal units (RTUs) by sending a short message to each RTU to inquire whether the RTU has anything to report. If it does, the RTU will send a message back to the master, and the data received will be put into the memory of the computer. If required, a control message will be sent to the RTU, and an alarm or message will be printed on the master typewriter and displayed on the cathode-ray-tube screen. In most systems the scan cycle, that is, the scan of all RTUs in the system, will be completed in approximately 2 s. However, in the event of trouble at a remote station, a message will be sent from the remote unit to the master, and the normal scan will be interrupted long enough for the master to receive the message and provide an alarm so that the operator can take immediate action, or, in some cases, so that the master unit can automatically perform predetermined control actions. In any event, in most cases, the status of all stations equipped with RTUs can be monitored every few seconds, providing the operators at the control center with up-to-date reviews of system conditions.

Almost all modern supervisory control systems are computer-based; that is, the master unit consists of a digital computer with the input and output equipment needed to transmit control messages to the remote units and to receive information from them. The received information is displayed on cathode-ray tubes (CRTs) and/or printed on electric typewriters for permanent records. CRTs can also display graphic information, such as one-line diagrams of the remote stations. In many control centers, overall system status is also shown on wall diagrams, which are kept current with existing conditions by data from the remote terminals.

Because of the wide acceptance of computer-based SCADA systems, which have made previous systems relatively obsolete, only computer-based systems are discussed in this chapter.

COMMUNICATIONS FOR SCADA SYSTEMS

As has been noted, SCADA systems consist of a master station, remote terminals (RTUs), and some communication links between the master and the remote units. The communication links can be wire circuits, microwave channels, or power-line carrier channels. Any communication circuit that provides an adequate signal-to-noise ratio and has a bandwidth capable of passing the data signals at the rate at which they are transmitted can be used.

Higher signaling speeds require increased bandwidths of the data communication channel used. In most applications, a normal telephone voice channel of about 400- to 3400-Hz bandwidth is satisfactory. For slow-speed data transmission, a narrow bandwidth, usually located above the highest voice frequency, can be used, however. In such cases, the voice band is restricted to about 400 to 2200 Hz, and the data are transmitted from 2200 to 3400 Hz. This type of operation is called "speech-plus," and it provides for both voice and data communication on a single voice channel, with a somewhat degraded voice channel.

It should be stressed that communication is of primary importance for a SCADA system. Poor communication results in errors or lost messages. A system cannot function properly without reliable and adequate communication channels.

CONFIGURATION OF SCADA SYSTEMS

As previously mentioned, SCADA systems consist of a master unit and RTUs. Various configurations can be used, and the configuration selected is determined by the requirements of the system, availability of communication channels, and cost factors. Figure 8-1 shows various possible configurations of SCADA systems.

Reliability of SCADA systems can also be increased by providing alternative communication channels, so that in the event of failure of a communication circuit, the affected RTUs can be connected to another communication circuit, usually automatically. It should be apparent that the overall reliability of a system can be no better than that of the communications between the master and the remote units, and because communication channels are exposed to various hazards, they are usually the least reliable part of such a system.

SUPERVISORY MASTER UNITS

The master unit of a supervisory system is the heart of the system. All operator-initiated operations of an RTU are made through the master unit and are reported back to the master from the RTUs. As previously noted, modern supervisory master units consist of a digital computer and equipment to permit communications between the master and the RTUs. Such equipment includes modems (modem is a contraction of

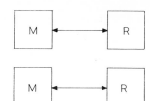

figure 8-1(a) Block diagram for a one-for-one SCADA system configuration, with a master unit for each remote unit.

(a)

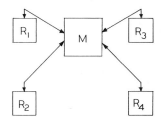

figure 8-1(b) Block diagram for a star or hub SCADA system configuration, with one master unit for several remotes, but with only one remote on each communication circuit.

(b)

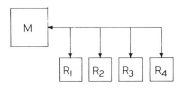

figure 8-1(c) Block diagram for a party-line SCADA system configuration, with several remote units on a single communication circuit.

(c)

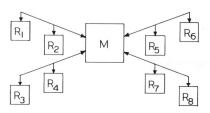

figure 8-1(d) Block diagram for a SCADA system configuration combining the characteristics of the star (Fig. 8-1b) and party-line (Fig. 8-1c) configurations, with one master and several remotes on each of various communication circuits.

(d)

modulator-demodulator) to convert the digital pulses used by the computer to a form that can be transmitted and received from the RTUs. Signals between the master and the RTUs are usually tones of audio frequency, and the messages are normally transmitted by frequency-shift techniques, as described in Chap. 7, "Telemetering Methods." Frequency-shift signaling is desirable because of its relative immunity from noise interference.

In addition to the computer, peripheral equipment necessary for the proper operation of the system is provided. Such equipment consists of a control console, keyboards or other means of entering data and commands into the computer, and cathode-ray tubes (CRTs) and typewriters to provide the operator at the master station with written messages of actions performed by the master and of data sent back from the RTUs.

In addition, a simplified one-line diagram of the power system is normally supplied in the form of a large wall map that shows when stations are normal and when abnormal conditions exist.

Digital-to-analog converters are provided to convert the digital data message information (on such items as line current, bus voltage, power, and reactive flow) to analog form that can be used to supply indicating or recording instruments. The recording instruments provide the dispatcher with a visual representation of conditions on the system, and of each remote station when it is selected to be displayed at the operating center.

Most modern systems also provide graphic displays, that is, one-line diagrams on the CRT screen. With color CRTs different colors can be used to distinguish between voltage levels, and changes of color can be used to indicate whether a circuit breaker is open or closed. A flashing indication is automatically used for a device that has changed its state, as well as an audible alarm, to alert the operator to the condition. The flashing indication and the audible alarm stop when the operator acknowledges the alarm condition.

When an operation is performed from a SCADA master unit, every effort is made to ensure that the desired device is selected and that the correct operation is chosen. The operator at the master station follows a procedure called the "select before operate" method, which is given below:

- The operator selects the remote station.
- The remote station acknowledges that it has been selected.
- The operator selects the device to be operated.
- The remote unit acknowledges that the device has been selected.
- After receiving assurance that the desired operation has been acknowledged by the remote unit, the operator performs the operation.
- The remote unit then performs the operation and signals back to the master that it has been performed, by changing color on the CRT display and by a printed message on the logging typewriter.

Following such a procedure holds the probability of erroneous operations to a minimum.

SUPERVISORY REMOTE UNITS

The remote units of a supervisory system are located at selected stations, and are either wired to perform certain preselected functions or, in modern units, equipped with microcomputers which have memory and logic capabilities. RTUs with microcomputers, called "intelligent remotes," can perform some functions without the direction of the master unit. However, any operations performed are reported to the master on the next scan.

The RTUs can also drive a new element, the programmable controller (PC), a dedicated controller with memory and logic.

The remote units are also equipped with modems so that they can accept messages from the master and signal back to the master that messages have been received and the desired operations performed. Relays located in the RTUs are used to open or close the selected control circuits to the controlled equipment on command from the master unit, and to sense when an operation has been performed so that the RTU can signal back to the master that the desired operation has been completed.

Transducers in the remote units are used to convert such quantities as voltage, current, watts, and vars to direct current or voltage proportional to the measured quantity, and then by means of analog-to-digital (A/D) converters convert the quantity to digital form, used by the system for transmission from the remote to the master.

Status Indication

The status of various devices—that is, whether open or closed, in service or out of service—is indicated by relays controlled by the position of the monitored devices. By such means, the master unit can be informed of the status of each remote station on each scan of the RTUs.

Momentary-Change Detection and Sequence-of-Events Recording

RTUs equipped with adequate memory can also store information, limited by the amount of available memory, so that changes of status that might complete a full cycle between scans from the master unit can be stored and reported on the next scan of the remote. An example might be the automatic tripping and reclosure of a circuit breaker, where an open-close operation can be completed during the interval between the time that the remote is scanned and the next scan. In such a case, the fact that there has been an operation would be completely lost without momentary-change detection (MCD).

A more sophisticated system, called sequence-of-events recording, is used where several actions might take place in rapid sequence. For example, the operation of both phase-distance and ground-distance relays on a single fault can occur with very little time between the relay operations. Protection engineers would be interested in determining which set of relays operated first and whether the sequence was correct. With sufficient memory in the RTU, and with timing provided by an internal clock or from the master, the sequence and times of occurrence of the relay actions can be stored in the RTU memory and reported back to the master on the next scan. Events that occur only a few milliseconds apart can be identified in their correct sequence. Sequence-of-events recording has become quite generally used in recent years.

With both MCD and sequence-of-events recording, the information stored in the memory of an RTU is erased after it has been reported to the master so that the RTU is prepared for the next event or sequence of events that should be reported.

OPERATIONS LOGGING WITH SCADA SYSTEMS

In addition to the control functions that have been briefly described and the information concerning quantities telemetered back from the RTUs to the master, a SCADA system can provide complete logs of the operation and status of the system or portion of the system under its surveillance.

Modern master units are equipped with computers which have both core and bulk memory, usually in the form of magnetic disks, and consequently are capable of storing a great deal of information that can be called out of the memory when desired. The computers also have internal clocks so that they can retrieve stored information at preselected times, such as each hour, and print logs of events that have occurred in the desired time period. Such logs could include the time of event occurrence, the status of all circuit breakers (that is, open or closed), totals of energy consumed or generated during the period, line and equipment loadings, etc. These are routine logs. In addition, emergency logs can be provided, so that immediately following any trouble the equipment involved, the time of occurrence, and other desired information can be made available to the operators. Other types of information, such as the occurrence of an overload condition, can be recognized, and an alarm provided before damage occurs.

Logs are normally printed by a typewriter or a line printer so that a permanent record is available. However, the information may also be displayed on the CRT, giving the operator immediate access to the information when desired.

The master unit computers are also capable of performing calculations and can total selected quantities over a day or other desired period, relieving the operators of much tedious work that was necessary when all such calculations were done manually. As a result, the system can be operated with greater reliability and safety.

ADDITIONAL SCADA SYSTEM APPLICATIONS

In addition to the remote supervisory control, status monitoring, and logging capabilities of a SCADA system, already described, various other programs can be incorporated in such systems to improve operations and minimize the manual effort required of power system operators. Some of these functions are briefly described below.

Automatic Generation Control

The bases for power system control were described in Chap. 5. Modern power system control normally makes use of control systems that are responsive to frequency variations, cost factors, transmission losses, etc., and that send corrective control pulses to generating units under the control of the system. Early systems were of the analog type and were called load-frequency-control (LFC) systems. Such systems could control only frequency, but that, of course, was a great step forward over governor control only. With the advent of the digital computer and its application to control systems, the control possibilities were greatly increased. A simplified block diagram of a computer-based control system is shown in Fig. 8-2a. These systems provide frequency control with transmission loss considerations, control of interconnected tie lines to other systems, and economic operation as described in Chap. 4, as well as other capabilities which will be described later.

Some power systems use a separate computer for system control functions. However, as modern SCADA systems are computer-based, it is possible, and is becoming almost universal practice, to use the SCADA system computer to provide the automatic generation control functions as well (see Fig. 8-2b). Of course, the computer must have more capability than would be needed for supervisory control only. Because of the rapid developments in computer technology in the past few years, computers providing increased capabilities at reduced cost are available for SCADA systems and other control functions and are receiving wide acceptance and application.

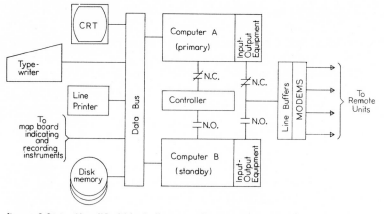

figure 8-2(a)   Simplified block diagram of a computer supervisory master unit. As shown in the diagram, there are two computers, one as the primary and the second as the standby. With failure of the primary computer, transfer can be made to the standby computer to minimize the possibility of loss of the system. Transfer can be made automatically or manually; however, most modern systems use automatic transfer. The standby computer is normally kept updated so that there will be no significant loss of data or control.

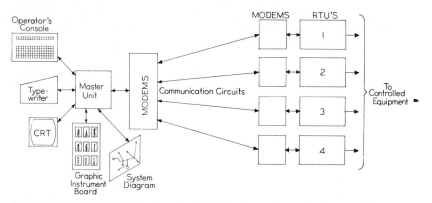

figure 8-2(b)   Simplified diagram of a SCADA system, showing the overall layout of such a system.

Security Monitoring

When sufficient information on the conditions existing on a power system is made available to a SCADA system computer, the computer can be programmed to check the limits of loading and other quantities in order to determine whether the system is at or near an emergency state. It can then cause an alarm to be sounded so that the system operator can take timely steps to eliminate such conditions and restore the system to a normal status.

Static-State Estimation

Some efforts have been made to develop procedures for estimating the state of a system, including calculated values of nontelemetered information which can be calculated from other available data; the result is an estimate of the security of the system.

Steady-State Security Analysis

Security analysis is used to determine whether a power system is secure and could stand the outage of certain lines or equipment without a system emergency. It is also used to determine a strategy for corrective action to restore the system to a normal security condition.

Online Load Flow

When sufficient information is telemetered to the master unit, a load-flow program can be developed using actual operating data. Such a program can predict loadings of lines and stations under selected future conditions. This type of program is also used to calculate the penalty factors (transmission-loss factors) needed for economic operation of the system, as was discussed in Chap. 4. It should be pointed out that this type of load-flow program does not take the place of the engineering-type load-flow studies used for the design of power systems, but it does provide an important working tool for the system operators.

SCADA SYSTEM RELIABILITY FACTORS

The reliability of a SCADA system is very important, and several means are used to ensure maximum reliability for such systems. Most master units are dual computers, with one as a primary unit and the other on standby to take immediate control, usually automatically, if the primary unit should fail. Also uninterruptible power supplies are used so that loss of commercial ac supply will not interrupt the operation of the system.

The security of the messages between the master and the remote units is maintained at a maximum by special coding checks that will detect errors in transmission of a message and have it retransmitted.

SUMMARY

This section has presented a brief description of SCADA systems and has outlined some of the capabilities of such systems over and above supervisory control. The application of digital computers to such systems has provided very powerful tools for system dispatchers, so that they can be kept aware of system status and can also be provided with automatic logging, automatic generation control, and other operations considerations.

Such systems have greatly increased the ability of system operators to maintain complete and timely information on system conditions and to rapidly take appropriate action during trouble periods.

PROBLEMS

1. Wired supervisory control systems are
    (a) Less expensive when many devices are to be controlled
    (b) More reliable than computer-based systems
    (c) Limited by cost and complexity when many devices are to be controlled
    (d) Immune from induced electrical interference
2. Electronic communication methods for supervisory control
    (a) Reduce the reliability of the system
    (b) Are more expensive than wired systems
    (c) Provide increased capability over wired systems
    (d) Can only be used with sequential scanning systems
3. Supervisory master units are normally installed at
    (a) Remote stations
    (b) Control centers
    (c) Centers with computer capability
    (d) Thermal power plants
4. Supervisory remotes (RTUs)
    (a) Report all conditions at their stations on each scan
    (b) Initiate the control actions to be performed
    (c) Frequently use exception reporting techniques to reduce data transfer
    (d) Can report only status information to the master station
5. Computer master units with dual computers require
    (a) Manual transfer to the backup computer in the event of failure of the primary computer
    (b) Battery-operated power supplies
    (c) Duplicate communication circuits to each RTU
    (d) Some means of transfer, either manual or automatic, to the standby unit
6. Communications for SCADA systems are
    (a) Not affected by signal-to-noise ratios
    (b) Required to be by microwave channels
    (c) Important to ensure reliability of the system
    (d) Always of the "speech-plus" type
7. SCADA systems require
    (a) A separate communication channel for each RTU
    (b) Continuous attendance at each remote station
    (c) Frequency-shift data channels to the RTUs
    (d) A means of entering and retrieving data at the master unit
8. In the star or hub configuration:
    (a) A separate master is required for each remote unit.
    (b) One master can serve several remote units.
    (c) There can be several remote units on the same communication circuit.
    (d) The system is less expensive than with a party-line configuration.

9. Digital-to-analog converters
  (a) Convert quantities such as voltage and current readings to a form that can be transmitted on supervisory data channels
  (b) Provide graphic displays
  (c) Convert digital data to analog form to supply indicating or recording instruments
  (d) Are used only at RTUs

10. Select before operate procedure refers to
  (a) The selection by the RTU of the function to be performed
  (b) A method whereby the operator checks that the correct devices have been selected before performing an operation
  (c) The master unit automatically selecting the operation to be performed
  (d) The display of single-line diagrams on the CRT

11. Transducers are used to
  (a) Convert digital quantities to analog values
  (b) Transmit messages between the master and the RTUs
  (c) Convert voltage readings to watts
  (d) Convert quantities such as volts, amperes, watts, and vars to currents or voltages proportional to the measured quantity

12. Momentary-change detection (MCD) is
  (a) A means of storing information on changes of status at a remote station between scans by the master unit
  (b) A method of providing read-only memory (ROM) in an RTU
  (c) Never used with automatic reclosing circuit breakers
  (d) A method of distinguishing between events that occur in rapid sequence

13. Logging with SCADA systems
  (a) Increases the amount of manual effort required of the operators
  (b) Can provide both routine and emergency logs
  (c) Can only be accomplished by operator request
  (d) Refers only to items such as equipment overloads

14. Automatic generation control (AGC)
  (a) Requires a separate computer
  (b) Can be provided by a special program in the SCADA master unit computer
  (c) Refers to the paralleling and separating of generating units
  (d) Cannot include transmission losses

15. Security monitoring with a SCADA system
  (a) Reports conditions on a power system after an event has occurred
  (b) Checks limits and other quantities that determine whether or not a system is at or near an emergency condition
  (c) Provides only audible alarms of emergency conditions
  (d) Checks the status of all circuit breakers at remote stations

16. Online load-flow programs of a SCADA system
  (a) Use actual operating data
  (b) Use calculated values of loading of lines and equipment
  (c) Cannot provide transmission-loss factors
  (d) Are used only in system design functions

17. SCADA system reliability is ensured by
  (a) Providing air-conditioned locations for the master units
  (b) The use of a computer with adequate memory capacity
  (c) The use of dual computers and uninterruptible power supplies
  (d) Retransmitting all data messages

9

Power System Reliability Factors

INTRODUCTION

Providing reliable service from a power system is one of the most important responsibilities of power system operators. This factor receives a great deal of attention in the design and construction of power system equipment and transmission and distribution lines.

Generation and substation equipment is carefully engineered to give many years of reliable service and has design provisions to withstand transient overvoltages due to lightning or switching surges. Equipment is also designed to withstand the mechanical and electrical stresses that may result when it is subjected to high fault currents.

System design provides for sufficient capacity in lines and station equipment so that equipment failure will not ordinarily result in customer load being interrupted in the event of loss of a line, transformer bank, circuit-breaker bushing, or similar trouble.

A commonly used design criterion is to provide facilities and capacity to withstand one foreseeable contingency, such as the loss of one line, one transformer, or other credible occurrence. Usually system design does not provide for second or greater contingencies because of the excessive cost and the reduced probability of two events occurring simultaneously.

After a power system is designed and constructed, it is the responsibility of the system operators to operate the system so that the design limits are not exceeded, to be alert to conditions that may exist that could affect reliability, and to be ready to take action to prevent hazardous situations from developing. Following trouble, when service is lost or equipment is unavailable, the system operator should proceed to restore the system to as near normal operation as possible so that its reliability is maintained at the highest possible level.

FACTORS AFFECTING POWER SYSTEM RELIABILITY

Some of the major factors that affect power system reliability are:

1. Reserve capacity
2. Adequate transmission and station capability
3. The ability to match load with generation
4. Prompt disconnecting of faulted lines or equipment and restoration of facilities
5. The ability to restart generation equipment
6. The ability to operate equipment such as power circuit breakers without dependence on power system energy
7. The ability to provide alternative arrangements of lines or station equipment to restore unfaulted equipment to service promptly
8. Adequate and reliable interconnections with other systems
9. Reliable indication of system conditions and communications with key generation and transmission stations

The above list of items is by no means complete, but it covers major categories that must be reviewed in order to ensure reliability of power system operation. Some items listed are determined by design and are not within the control of system operators. The following discussion will attempt to point out factors over which system operators can exert control in order that maximum reliability can be achieved with the available facilities.

SPINNING RESERVE

Generating capacity that is on the line and that is in excess of the load on the system is called spinning reserve. Adequate spinning reserve is probably the primary security factor in power system operation.

The amount of spinning reserve that a system desires to carry is a policy decision based on risks and economics. Once the spinning-reserve policy has been established, it is the duty of system operators to attempt to meet the criterion each day so that the system will not be jeopardized by inadequate reserve. Because of the no-load fuel costs incurred by excessive reserve, system operators also should see that excessive reserve is not carried.

The amount of spinning reserve can be expressed as a percentage of the daily peak load or be based on the risks of loss of generating capacity that actually exist on the system. The determination of spinning reserve as a percentage of the daily peak load leaves much to be desired, since it may not take into consideration the actual risks that exist on a system. Furthermore, particularly with thermal generating equipment, units usually require several hours from the time that they are ordered to be placed in service before they are actually available. Consequently, it is necessary to make estimates of load, which may be somewhat in error. If load is underestimated, the percentage of spinning reserve at peak time may actually be less than that required by the criteria. Interconnection agreements sometimes provide penalties for inadequate spinning reserve.

Probably a more realistic method of specifying spinning-reserve criteria is to base them on risks, along with allowances for forecast error and regulating requirements. Elements of risk include the load on the most heavily loaded unit or the amount of power being imported into the system if there are interconnecting tie lines. In addition to the risk due to unit or tie-line load, allowance is made for forecast error and for regulation error. These factors are usually 2 to 3 percent each. In some cases another factor is added, an arbitrarily determined amount which takes into consideration abnormal system arrangements or other conditions which might result in a higher-than-normal risk.

An example of calculating required spinning reserve on the basis just outlined is illustrated in Fig. 9-1 for system B.

Another factor connected with spinning reserve that should be under the continuous scrutiny of the system operator is its location and makeup. In the event of loss of a large, heavily loaded unit, frequency will drop. The amount of frequency sag will depend on the proportion of total available generation lost, including that on interconnected systems. It is desirable to restore frequency to normal as soon as possible, and in the event of interconnected operation to return tie lines to normal schedule as soon as possible so that instability or overloads will not occur, inadvertent energy accumulations will not become excessive, and the effect of the trouble on the interconnected area will be minimized.

Generating units have limits on the rate at which they can respond

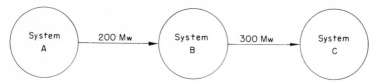

figure 9-1(a)   Assume a system peak load of 4000 MW with the most heavily loaded unit 350 MW, a 2 percent forecast error and a 3 percent regulation error, and no abnormal system arrangement.

$$
\begin{aligned}
\text{Most heavily loaded unit} \ &= \ 350 \ \text{MW} \\
\text{Forecast error} \ &= \quad 80 \ \text{MW} \ (0.02 \times 4000) \\
\text{Regulation error} \ &= \ 120 \ \text{MW} \ (0.03 \times 4000) \\
\text{Contingency factor} \ &= \quad \underline{0} \\
\text{Required spinning reserve} \ &= \ 550 \ \text{MW}
\end{aligned}
$$

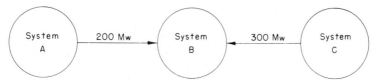

figure 9-1(b)   Assume the same system and largest unit loads as in *A* and an arbitrary contingency factor due to abnormal system arrangement of 100 MW.

$$
\begin{aligned}
\text{Net imported power} \ &= \ 500 \ \text{MW} \\
\text{Forecast error} \ &= \quad 80 \ \text{MW} \\
\text{Regulation error} \ &= \ 120 \ \text{MW} \\
\text{Contingency factor} \ &= \ \underline{100 \ \text{MW}} \\
\text{Total spinning reserve} \ &= \ 800 \ \text{MW}
\end{aligned}
$$

when a load pickup is required. With hydro units, rates of pickup are usually limited by the rate at which water can be accelerated in the penstocks, and with thermal units, after the initial energy stored in the boilers is used, by the rate at which steam can be produced to sustain a load pickup.

It is possible to determine the percentage of unloaded capacity that can be picked up by the various generating units in time categories such as 5 s, 10 s, 30 s, 1 min, 5 min, etc. With such a determination of spinning reserve, it is possible to predict reasonably well how a system will respond to trouble resulting in a frequency drop.

With large-scale interconnections, the loss of even a large, heavily loaded generating unit will not produce a significant frequency drop. In such cases the power angle of the system losing generation retards, and there is an immediate flow of power from other interconnected systems into the system with deficient generation. Since there is little or no frequency drop, the deviation of tie lines from normal schedules is the only indication of abnormal conditions. Tie-line-load frequency-control signals or telephonic orders to plants with reserve capacity are necessary to restore tie-line schedules to normal. In such cases spinning-reserve response is somewhat slower than when a significant frequency drop occurs.

A very important factor in maintaining proper spinning reserve is to have the reserve distributed over several units throughout the system. If most or all of the reserve is on one large unit, the total response is limited to the rate at which that one unit can pick up load. When reserve is divided among several units, each can provide its share in restoring system conditions to normal, and the possibility of instability, tie-line tripping, or line and equipment overloads is reduced.

To summarize, adequate spinning reserve is one of the major factors in maintaining power system security. The amount of spinning reserve to be carried is based on an evaluation of risk and is a management policy decision. After the policy has been decided, it is the responsibility of the system operators to ensure that the policy is met and that allocation of reserve among available units is such that proper response to loss of generation or tie lines will be achieved.

TRANSMISSION AND STATION CAPABILITY

The power-handling capabilities of transmission lines and substation equipment are design factors, and are not under the control of system operators. However, after lines and equipment are installed and in service, the system operator is in a position to see that capabilities are not exceeded in normal operation. By frequently monitoring load and voltage conditions at various locations in a system, the operator can be kept aware of conditions and can adjust generation or alter arrangements to prevent overload conditions from occurring.

System operators should be familiar with normal and overload ratings of facilities under their jurisdiction. Some equipment, particularly transformers, can be operated at greater than nameplate rating for limited periods of time without damage. Generating equipment ratings are established by the manufacturer and confirmed by operational tests following installation.

EFFECTS OF TEMPERATURE ON EQUIPMENT

The limiting loading factor on all electrical equipment is temperature rise.

Generating and substation equipment maximum operating temperatures are specified by manufacturers and by information provided by system engineering and operating organizations. If the ambient air temperatures are low, it is usually possible to load equipment to higher loads than with high ambients. With modern thermal generating equipment, it is possible to gain appreciable capacity temporarily at the expense of efficiency by cutting out feedwater heaters. In emergency conditions the additional capacity available by this means may prevent overloads on other equipment or emergency load shedding. Increasing pressure in thermal unit boilers, within limits, can also be used to temporarily increase capacity.

POWER-FACTOR CONSIDERATIONS

Another factor that should be under the constant surveillance of the system operator is the power factor of generating equipment. If a unit is supplying a relatively large var output, its total rating may be exceeded even though the MW load is below rating. When leading var are being supplied, there is an increased possibility of heating the end laminations of generator armatures. Temperature-sensing devices such as resistance thermal devices (RTD) or thermocouples are usually provided to monitor such conditions.

TRANSMISSION-LINE RATINGS

Transmission-line ratings are determined by conductor type, size, and line length. For short lines, only the type and size of conductor are of importance because the electrical phase shift of the line under heavily loaded conditions is not sufficient to cause a stability problem. On such lines the thermal rating of the conductor is the limiting factor.

Usually lines where thermal capacity limits capability are given summer and winter ratings. The summer ratings are somewhat lower than the winter ratings because of higher summer ambient temperatures.

On long lines, the rating may be determined by stability limitations rather than by conductor thermal capability. On such lines the stability limits are reached before the conductor current reaches its maximum limit.

A system operator who is familiar with line and station capabilities can take necessary action during both normal and trouble conditions to ensure that capabilities are not exceeded or, if necessary, to modify switching arrangements or shed load to assist in providing maximum service reliability.

MATCHING GENERATION WITH LOAD

When a power system is operating at normal frequency, with tie lines to other systems carrying scheduled loads, generation and load are matched. Any increase or decrease in load must be followed by a corresponding change in generation to match the new load condition.

The system operator is provided with various indicating devices, including system frequency, telemetered tie-line flows, and area control error, so that the operator can be kept continuously aware of these factors. Matching generation and load is the basic responsibility of the system operator. Devices such as load-frequency-control and automatic-dispatch equipment are provided to help match generation and load. It is only after this is accomplished that economic loading of generation is effected.

To make sure that there is always sufficient generation capability to carry any expected load, with provision for the loss of a major generating unit, spinning-reserve capacity is carried, as previously discussed. However, if serious troubles occur, such as the loss of all tie lines or a total generating station from bus trouble, the remaining generation may not be sufficient to carry the load on the system.

When generation is inadequate, system frequency will decline. It is of extreme importance that frequency be prevented from continuing to decline. In thermal plants particularly, auxiliary devices such as boiler feed pumps, draft fans, etc., must operate at or near normal speed and voltage to function properly. A frequency drop of more than a few hertz (5 or 6) may cause a loss of the plant auxiliaries and result in complete loss of the plant, further reducing the available generation and increasing the possibility of a total system collapse.

If frequency decline continues, it is necessary to match load with the available generation. This can be done manually by dropping customer load promptly in sufficient quantity to arrest the frequency decay and start to restore frequency to normal.

Manual load shedding leaves much to be desired, since there usually is very little time for the system operator to assess the situation and take the proper corrective action. As a consequence, it is common practice to install underfrequency relays to disconnect load automatically in amounts sufficient to match the remaining load with available generation. In developing load-shedding programs, it is customary to drop load in increments with declining frequency. For example, if the credible incident could result in a loss of 30 percent of the generation, it might be decided to drop 35 percent of the load in stages, with 5 percent of load being dropped at 59 Hz and additional amounts dropped if frequency continues to decline, until the total 35 percent is dropped at some preselected frequency, such as 58 Hz.

Load-shedding programs vary among systems, but all are planned so that the maximum load is dropped before frequency is reduced to such an extent that plant auxiliaries are lost and a total system collapse occurs.

Another factor that can be used to minimize the possibility of system collapse with declining frequency is to open tie lines at some predetermined frequency if power is being exported from the system. If power is being imported, opening the ties will worsen the situation. Tie-line opening in time of trouble should usually be a last resort and should be instituted only if power is flowing out of the system with continuing decline of frequency. Interconnection agreements and operating orders issued by various systems usually cover this procedure, and it is also covered in North American Power Systems Interconnection Committee (NAPSIC) Guide 9. In most cases, interconnections will serve to stabilize and limit frequency decay.

A more sophisticated procedure by which the rate of frequency decay is used to determine the amount of load to be shed can be applied. Relays that are sensitive to rate of frequency decay have been developed, and these relays can supply frequency information to a central control computer, which can analyze frequency variations and send out control impulses to shed load when conditions require it.

As frequency recovers, load is reconnected in amounts that match the available generation. In some cases this is done automatically by relay, and in others manually. In any event, it is much more desirable to interrupt a portion of a system's load for a short period of time than to permit frequency decline until the system collapses.

Matching generation with load also is necessary in the event that a large block of load is lost, or when tie lines which are exporting power relay. In such circumstances, load will be less than generation, and frequency will rise. The amount of frequency rise must be limited because it is accompanied by voltage rise and may damage customer equipment if it is not limited. Overfrequency is less hazardous to a power system than low frequency. As frequency rises, the additional torque that is required from the prime movers tends to limit frequency rise. However, the total rise in frequency should be limited to some predetermined level. This is usually done manually by tripping generation if governor and automatic control action is not adequate.

The important principle to be remembered is that in power system operation, generation and load must always be matched. Any mismatch results in a variation of frequency from normal or results in deviations of tie lines from schedule. Inadequate generation results in reduced frequency and overgeneration in overfrequency.

It is important to prevent serious frequency decline so that plant auxiliaries will not be lost, causing further mismatch of generation and load, which can result in a total system collapse. Present practice is increasingly trending toward application of underfrequency relays to shed load during periods of serious frequency decline. Upon restoration of frequency, load can be automatically or manually restored.

In interconnected operation the tie lines will ordinarily help maintain frequency. However, if power is being exported at a time when frequency is declining, conditions can be improved by opening the ties, but this should be done only as a last resort, as it will make an adjacent area that is already deficient more deficient.

For overfrequency operation resulting from loss of load, it may be necessary to drop generation to restore the match between generation and load.

The maintenance of the match between generation and load is a primary and continuing responsibility of the system operator.

DISCONNECTING FAULTED LINES OR
EQUIPMENT AND RESTORATION OF FACILITIES

One important measure in maintaining power system security is the rapid disconnection of lines or equipment that are in trouble. Because rapid action is necessary, automatic devices usually are relied upon instead of manual operations.

The design and setting of protective relay systems is a function of system engineering and operating staffs and is not ordinarily a matter over which system operators have any control. However, system operators should be aware of the protective devices at important locations in the power system and their expected performance.

Knowledge of the type of protective devices in use and the portions of lines and equipment protected is of value in determining the nature and extent of trouble after a relay operation. Such information should provide the operator with clues on how to proceed in restoring the system to normal, or as nearly so as possible, in minimum time.

Many types of troubles, such as insulator flashovers on transmission lines, often are only momentary. The protection system design therefore frequently provides for automatic reclosure following such incidents. On the other hand, transformer bank differential or generator-elevated neutral relay operations usually indicate more serious troubles.

As a guide to system operators, procedures are normally provided which outline the steps to be taken after different types of relay operations or after unsuccessful reclosure tests. System operators should be thoroughly familiar with such system policies in order to restore the system to as nearly normal as possible with minimum delay.

In the event that automatic reclosure is unsuccessful, or after relay action has occurred that indicates equipment trouble, the equipment should be disconnected and cleared for repairs. Alternative line or station bus arrangements should be effected in order to restore interrupted load or generating equipment to service, so that normal loads will be served and generation margins restored.

RESTARTING GENERATION EQUIPMENT

After a generating unit relays or is lost to the system because of line or station trouble, it should be returned to service promptly in order to restore generation margin if there is no machine damage.

Normally, restarting a generating unit does not present particular problems, other than those of the routine procedures that must be followed in putting it into service. However, if there is no power available at the generating station, from the system, or from local startup or house units, there may be a long delay in getting the unit back into service.

DESIGN FACTORS AFFECTING RELIABILITY

A great deal of effort is expended in the design of generating stations to make them as reliable as possible and ensure maximum availability of the equipment for service. Some of the means used to enhance reliability are spare exciters, startup transformer banks, and battery or pneumatically operated control devices with capability of several operations from stored energy. In the event of a total system or area collapse, however, it may not be possible to restart a plant without sufficient electric power to supply auxiliaries.

In some thermal plants small "house" units are provided which will separate from the system on serious frequency drops and supply station auxiliaries, such as feed pumps, draft fans, lubricating oil pumps, and other devices essential to plant operation, at normal frequency and voltage. Such plants are normally capable of restarting with a minimum of difficulty after a complete separation or system collapse.

In order to increase system reliability, many installations of diesel or gas turbine generators of sufficient capacity to provide starting power are being made. Such startup units are capable of starting rapidly with only a battery or compressed air source, which is supplied as a part of the installation.

Hydro plants are ordinarily less complex than thermal plants, and if bearing oil supplies and control power are available, they can be started and placed in service in very short periods of time.

The features of plants on a system are the result of design considerations that are not within the province of the system operator. However, the operator should be aware of the capabilities for startup after a total shutdown as well as the normal operating capabilities of the various plants for which he or she is responsible. Some of the factors of which the system operator should be informed are:

1. Availability of startup power, and whether by house unit, diesel, gas turbine, or other source

2. Transmission sources to the stations in the event that it is necessary to bring power back to the station for starting

3. Sources within the system that can be used for starting other plants

4. Switching procedures to route startup power to stations that require outside or system power for startup

Most of the above information is usually available in written emergency procedures, which system operators should understand thoroughly so that facilities can be restored to service and normal operations resumed as soon as possible in the event of major system shutdowns.

The foregoing is by no means a complete discussion of all the problems of restarting generating equipment or of emergency procedures. These will vary for each installation and system, and must be covered by detailed information and procedures developed by the engineering and operating departments of each system. However, the subject is one of paramount importance in making it possible to provide maximum service under all conditions.

OPERATING EQUIPMENT WITHOUT NORMAL ENERGY SOURCES

During power system emergencies it may be necessary to operate equipment such as power circuit breakers or motor-operated air switches when normal sources of energy for operating such devices are not available.

Power circuit breakers, whether oil, air, or gas types, are provided with mechanisms to open or close the breakers as desired. Solenoids, pneumatic devices, or operators using energy stored in compressed springs are used for this purpose. In order to make these devices independent of the system power, station batteries of capacity sufficient to provide energy for several open and close operations are ordinarily provided.

It is possible that station battery sources may be lost because of battery cable failure or other cause, and it may be critically important to operate circuit breakers during the period of battery failure. Usually emergency means of operating such devices can be developed. Some spring-operated breaker mechanisms can be hand cranked to compress the operating spring so that the breaker can be closed.

With pneumatically operated breakers, when there is no compressed air available, it is possible to use a nitrogen bottle, temporarily connected to the air system, to operate the breakers.

Motor-operated air switches can usually be manually operated, either directly or by winding a spring operator.

The important factor is that even in the event of loss of normal sources of energy for operation of devices in a power system, methods can frequently be devised to permit operation during emergencies. System operators who are aware of possible emergency procedures can direct restoration of service with much less delay than if they were to wait for repairs to be made to damaged equipment or for system conditions to return to normal.

ALTERNATIVE ARRANGEMENTS

Normal arrangements of transmission and distribution lines and the configuration of lines to station buses permit proper division of load, ensuring minimum risk in the event of bus or transformer bank failure and proper relay action.

During emergencies, system operators are frequently required to devise alternative arrangements of line and station equipment in order to restore service with a minimum of delay. Common procedures are to parallel lines over an auxiliary bus or to make use of bus parallel breakers to replace a breaker that is damaged or out of service for work.

In some cases, in the event of a permanent fault on a line, the line can be sectionalized or jumpers opened at dead-end structures to restore at least a portion of the service until repairs can be made and the system returned to normal.

It is not possible to cover all the alternatives in this brief discussion, but system operators can review potential troubles and develop procedures to be used in such cases. It is common practice on most power systems to prepare standard switching procedures for various contingencies. However, it is never possible to foresee all possible contingencies, and thus intimate knowledge of the system will help the system operator to devise emergency arrangements when necessity arises and no prepared procedure is available.

INTERCONNECTIONS WITH OTHER SYSTEMS

Interconnection with other systems can be of significant assistance to a power system during times of trouble. In the event of loss of a large block of generation, energy will flow from surrounding systems to the system with deficient generation. The ability to provide mutual standby is one of the major incentives for power system interconnection.

In most cases, with moderate loss of generation, interconnection of systems will reduce the amount of frequency drop because the percentage of total interconnected capacity lost is less than if a system operates isolated from other systems.

Serious disturbances can result in tie-line overloads, and in some cases instability, resulting in tie-line tripping, which may lead to cascading relay operation and area shutdowns more extensive than would have resulted with systems operating independently. Some of the measures previously discussed can minimize the possibility of such major pool outages.

Maintenance of adequate spinning reserve with capability of rapid response, load-matching-relay (underfrequency) installations, and proper tie-line-relay settings will ordinarily prevent disturbances from developing into area-wide shutdowns.

System operators are in a position to monitor tie-line flows and other conditions on their systems, and by proper surveillance and action in emergencies can prevent or minimize the effects of troubles. If trouble conditions persist, accompanied by declining frequency, it is desirable to open tie lines if power is outgoing and save at least a portion of the area. As previously mentioned, this procedure is well outlined in NAP-SIC Guide 9. By maintaining a portion of the area in service rather than permitting a total collapse, normal operation can be restored much more rapidly than would be possible following a total area collapse.

Interconnection agreements usually specify minimum spinning-reserve requirements and outline mutual standby conditions. Knowledge of system capabilities, tie-line ratings, spinning reserve available, and other operating features is important so that the system operator can take appropriate action during emergency conditions.

INDICATION OF SYSTEM CONDITIONS AND COMMUNICATION

Because of the nature of their work, system operators must rely on communication and signaling devices to keep them aware of conditions on their power systems. Many of the key locations, large generating plants, points of interconnection, and major switching stations are many miles from the system operator.

In order to provide the system operator with information on the status of the system, key information is remotely indicated in dispatching offices. Control channels and telephone circuits are provided for automatic or supervisory control of equipment and for telephone contact with operators at various stations throughout the system and between control centers of interconnected systems.

The reliability of telemeter, control, and voice channels is of major importance to system reliability. Various means of providing communication facilities are used. Channels are sometimes leased from common-carrier telephone companies, power-line carrier circuits are established on power-transmission circuits, and privately owned microwave systems are installed by the utilities.

In order to ensure reliability of communications, it is common practice to provide more than one communication path to major generation, switching, and tie points. Such alternative channels are usually routed over diverse paths so that the probability of concurrent failure is minimized.

The power supply to terminals and repeater stations of microwave and power-line carrier installations is made independent of system power, or an auxiliary power source is provided for service during a power system interruption.

When adequate care is used in the design and installation of communication and indication facilities, the system operator has reasonable assurance of being able to maintain contact with key locations at all times and will be able to take the necessary actions to obtain the best possible reliability from the system.

SUMMARY

From the preceding discussion it should be apparent that the highest importance is attached to power system reliability. Those responsible for the design and operation of power systems devote a great deal of thought to this subject and invest substantial sums to ensure reliability, including the reliability of communications. Modern SCADA systems frequently include security monitoring, state estimation, and online load-flow programs to provide system operators with indications of potential problems affecting the reliability of their systems. When provided with the facilities outlined in this section, it should be possible for system operators to perform their functions properly in maintaining safe and reliable operation of their systems.

PROBLEMS

1. One of the factors affecting power system reliability is
   (a) The nominal system transmission-voltage levels
   (b) The ratio of hydro to thermal generation
   (c) The available reserve capacity margin
2. Spinning reserve is generating capacity that
   (a) Is available by starting gas-turbine generating units
   (b) Is synchronized and online, with capacity in excess of existing loads
   (c) Has the lowest fuel cost
3. The spinning reserve to be carried on a system is
   (a) Determined by the system operator (dispatcher)
   (b) A policy matter
   (c) Determined by the amount of load on the system
4. The risks affecting spinning-reserve requirements of a system, in addition to the most heavily loaded units, include
   (a) The net amount of imported power
   (b) The net amount of exported power
   (c) The time of the daily peak load

5. Immediately following a loss of generation on a power system that is interconnected with other systems:
   (a) Frequency will drop on the system that has lost generation.
   (b) Governor action will increase load on other generating units.
   (c) Power will flow into the system through the tie lines from other systems.
6. Spinning reserve should
   (a) Be divided among units at various locations on a system
   (b) Be carried on the largest unit available
   (c) Be reduced during off-peak hours
7. The power-handling capabilities of transmission lines are
   (a) Determined by system operators (dispatchers)
   (b) Set by design factors
   (c) Controlled by load-frequency-control equipment
8. When ambient air temperatures are low, a transformer bank
   (a) Must be limited to a lower load than when air temperature is high
   (b) Secondary voltage is lower than when the air temperature is high
   (c) May be loaded to a higher load than when temperatures are high
9. The load-carrying capability of a long transmission line
   (a) Is always limited by the conductor size
   (b) May be limited by stability considerations
   (c) Is reduced at low ambient air temperatures
10. On an interconnected power system, matching of generation and load is indicated by
    (a) Normal frequency and power flowing in on the tie lines
    (b) Normal frequency and power flowing out on the tie lines
    (c) Normal frequency and tie lines at scheduled load
11. In the event of a major loss of generation in an interconnected area, a system operator should
    (a) Immediately open tie lines to other systems
    (b) Open tie lines if power is flowing out of the system
    (c) Open tie lines only as a last resort with power flowing out and declining frequency
12. When underfrequency relays are installed on a power system, they should be set so that:
    (a) No load is shed until the frequency declines to 56 Hz.
    (b) Increments of load are shed as frequency declines.
    (c) 30 percent of the load is shed at 59 Hz.
13. House units are sometimes provided at thermal power plants to
    (a) Provide for station and yard lights
    (b) Keep the station batteries charged
    (c) Supply power to plant auxiliaries
14. When a power system operates on an interconnected basis with other systems, loss of generation on the system will cause
    (a) A larger frequency drop than if the system operated isolated
    (b) A smaller frequency drop than if the system operated isolated
    (c) Power to flow out on the tie lines

10

Power System Protection

INTRODUCTION

The protection of power system lines and equipment is a specialized field, and in most systems the application of protective devices is handled by a department charged with this function. Brief mention of system protection was made in the section "System Reliability Factors." No attempt will be made here to cover this subject completely. The basic principles of protection will be explained in order to provide system operators with a reasonable knowledge of the protective system action that should occur under various system conditions. More complete information is available in engineering texts and manufacturers' publications.

In general terms, protective systems monitor system conditions such as current flow, current unbalance, voltage, equality of power in and out of a bus or transformer bank, currents on both ends of generator windings, temperature, and other quantities. If conditions are abnormal, the protective relays will sense the abnormality and close contacts in a battery-powered circuit. This circuit will provide energy for operating circuit breakers to disconnect the line or equipment from the power system.

General classes of protection are:

1. Overcurrent (nondirectional and directional)
2. Current balance
3. Over- and undervoltage
4. Distance [impedance, admittance (mho), etc.]
5. Transfer trip
6. Power
7. Differential
8. Phase balance
9. Phase comparison
10. Frequency (over and under)
11. Temperature

In addition to the above listing, there are many specialized relays for particular applications, and in many cases the protective installation may include several types of relays.

For power system applications, relays are usually designed to operate with relatively low currents and voltages. These quantities are of the order of 5 to 10 A in the current windings and approximately 115 V on potential (voltage) windings. Because the voltages and currents found in power systems are much higher than can be directly applied to the relays, potential and current transformers of known winding ratios are used to reduce these quantities to ranges that can be applied to the relays. Potential and current transformers were discussed in the section "Energy Accounting."

RELAYING PRINCIPLES

In any electromagnet the magnetic force (field) is proportional to the current through the coil, the number of turns in the coil, and the magnetic characteristics of the material used for the core. Consequently, if the same magnetic core is used, a coil with 10 turns and 100 A flowing in it will produce the same magnetic field as a coil with 1000 turns and 1 A flowing in it. If many turns of small wire are used, an electromagnet may be used across a line and be sensitive to voltage changes. With a few turns of large conductor it can be placed in series in the circuit and be sensitive to current changes. Most relays make use of electromagnets, although there are now many relays available using transistors and logic circuits which will perform the same functions as electromechanical relays. This discussion will be confined to electromechanical relays because the principles of protection are the same in both electromechanical and so-called solid-state relaying.

Probably the simplest type of relay is a solenoid (coil of wire) with a movable element of magnetic material (iron) inside the coil or adjacent to it. When current flows in the coil, a magnetic field is produced. If sufficient current flows, the movable element will be pulled up or drawn toward the coil. When electrical contacts, electrically insulated from the movable element, are attached, the contacts will move with the element, and can either close or open control circuits. The operating coil of such a relay can be connected in series with the monitored circuit to be operated by current flow, or connected across the circuit to be sensitive to voltage changes. Examples of these simple relays are shown in Fig. 10-1. They are frequently used for motor starting or for auxiliary applications in power system protection installations.

Relays of the types described above could be used for protection of power system circuits, but they have a serious deficiency in that they are not capable of providing time delays in their operation. When sufficient current flows in the relay coil, its contacts will close instantaneously. In fact, this type of device is used as the instantaneous element in more complex relay systems.

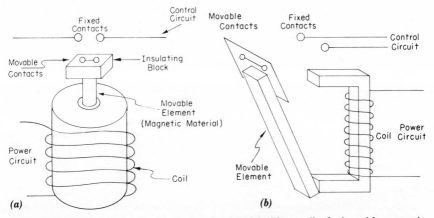

figure 10-1 Simple relays. *(a)* A plunger is placed inside a coil of wire with magnetic material below the center of the coil. When sufficient current flows in the coil, the plunger is drawn up and movable contacts bridge the fixed contacts to close the control circuit. *(b)* When sufficient current flows in the coil, the movable element is drawn toward the magnetic core and the movable contacts bridge the fixed contacts to close the control circuit. By placing the fixed contacts on the opposite side of the movable element, the control circuit will be normally closed and will be opened when the movable element is pulled up.

INVERSE TIME RELAYS

Usually it is desirable to incorporate time delays in power circuit protection, so that after a selected minimum current is reached, any increase in current will cause the time of relay operation to decrease as current increases. Relays have been developed that will provide contact closure in times inversely proportional to the square of the current in the circuit.

Inverse-time overcurrent relays permit short-term overloads of circuits, but for faults (short circuits) where excessive current flows, the relay contacts are closed to trip the circuit breakers in very short time intervals, so that damage to the line conductor or equipment is held to a minimum. In effect, inverse-time relays make use of motor action, where the torque of the motor element is proportional to the square of the current flowing in the circuit. Consequently, at currents only slightly above those required to cause the motor element to rotate, considerable time is required for it to rotate to the point at which the contacts are closed. As current increases, the rotating (rotor) element accelerates more rapidly, reducing the time required for the contacts to close.

The usual construction of an inverse-time overcurrent relay is somewhat similar to that of a watthour meter, which has previously been

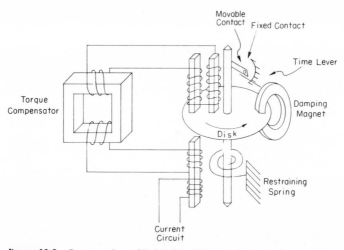

figure 10-2 Construction of induction-disk overcurrent relay. Current through the current coil of the lower pole produces a magnetic field, which induces eddy currents in the disk and also, by transformer action, into the upper poles. The eddy currents produce motor action, tending to cause the disk to revolve against the force of the restraining spring. The damping magnet also tends to prevent rapid acceleration of the disk. The moving contact can be moved toward or away from the fixed contact, changing the distance the disk rotates, which affects the time for the contacts to close.

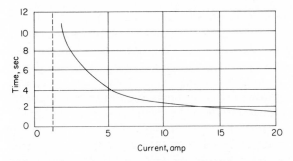

figure 10-3 Typical inverse-time characteristic curve of an induction overcurrent relay. In a practical application with a 100:5 current transformer, 100 A in the circuit would cause 5 A to flow in the relay current coils and tripping to occur in approximately 4 s; 200 A in the circuit would cause 10 A to flow in the current coils and the contacts to close (tripping) in slightly over 2 s. The time curve can be moved up or down by adjustment of the time lever (adjustable moving contact) shown in Fig. 10-2.

described. The rotating element consists of a nonmagnetic metallic disk, usually aluminum, with magnetic field coils placed near the disk. The construction of induction-disk overcurrent relays is shown in Fig. 10-2.

Representative time characteristics of induction overcurrent relays are shown in Fig. 10-3.

Induction overcurrent relays of the type described will operate whenever current in the circuit exceeds the minimum required to cause the relay disk to revolve; they are insensitive to the direction of power flow in the circuit.

DIRECTIONAL RELAYS

The next step in providing selectivity in protecting circuits is to make the protective system sensitive to the direction of power flow in the circuit. Sensitivity to direction of power flow is obtained by adding another element to a relay. The second element is similar in construction to the overcurrent disk element, but is provided with both current and potential (voltage) coils. When the power flows in one direction, the torque on this (directional) element will cause the disk to rotate in a direction that will close the contacts. When power flows in the opposite direction, the disk will attempt to rotate in the direction that will open the contacts further. (Rotation in the opening direction is prevented by mechanical stops on all except duodirectional and balance relays.) By connecting directional contacts to the overcurrent element, a relay sensitive to current and direction is obtained, as shown in Fig. 10-4.

Ordinarily, current directional relays are connected so that the directional contacts will close with power flowing away from the station bus, but remain open for power flowing into the station. By using this arrangement, protective relay operation can be confined to a line section that is actually in trouble without causing relay action on other lines, even though the current in them may be above the minimum required to cause the overcurrent contacts to close. An example of the application of directional overcurrent relays is shown in Fig. 10-5.

It can readily be seen that power will flow away from the buses at the remote stations supplying the lines to stations *A* and *B*. Consequently, the directional contacts on the relays protecting the remote ends of these lines will close. If relay current settings and time delays are properly coordinated, the relays nearest the fault will normally clear the trouble before the relays at more remote locations can operate, even though power flow is in the direction to permit tripping. However, in the event that relays in the faulted section fail to trip the circuit breakers, because

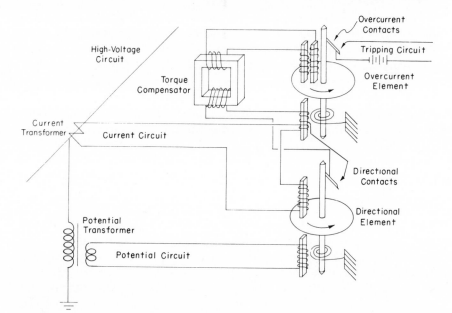

figure 10-4 Construction of directional overcurrent relay. If the directional element contacts are closed, the upper (overcurrent) element will rotate to close the contacts whenever current in the secondary of the current transformer circuit exceeds the minimum required to overcome the restraining force of the spring. The contacts of the lower (directional) element will close when polarities of the current and potential circuits are such as to cause rotation of the disk in the direction indicated. With reversed power flow, the torque on the directional element disk will tend to cause the disk to rotate to open the directional contacts. As a result, no torque will be produced to rotate the overcurrent disk, as the primary circuit of the torque compensator is open.

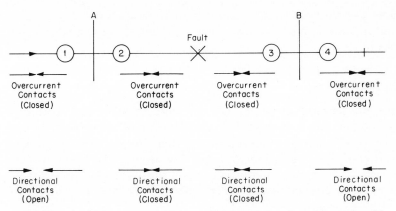

figure 10-5 Simplified diagram of directional overcurrent protection. (*Note:* For simplicity the current and potential circuits to the relays are omitted. They would be connected as shown in Fig. 10-4.) Assume a fault on the line between stations *A* and *B* at the point indicated. Power will flow into station *A* on the line from the left and into station *B* on the line from the right. In both cases the direction of flow is such that the directional elements of the relays on these lines will remain open. With sufficient current the overcurrent element contacts will close, but the tripping circuit will not be activated and breakers 1 and 4 will not trip. In the line between stations *A* and *B*, current will flow away from the stations toward the fault. Both the overcurrent and directional contacts will close on the relays for breakers 2 and 3, and when these breakers trip the line will be deenergized, clearing the fault from the system.

of contact or control circuit troubles or from mechanical malfunction of the circuit breakers, the remote breakers can operate as backup protection. In such a case a greater portion of the system is affected, but severe damage to lines or equipment can be minimized.

PARALLEL-LINE PROTECTION

In some cases two identical lines operate in parallel between stations. In this situation, under normal conditions, the current in the parallel lines will be equal. Current transformer connections are made so that the relays will maintain contacts in an open position when current equality exists and develop a torque to close contacts whenever currents are unequal. With such a protective installation, if one line has a fault, the currents will be unbalanced with the greater current being in the line with the fault. The balance contacts will close very rapidly to trip the faulted line.

By the use of a combination of balance relays and directional overcurrent relays on a pair of lines, if one line relays, the remaining line can be left in service with directional overcurrent protection until normal parallel operation can be restored.

Balance protection of parallel lines is being replaced with relay schemes that will provide high-speed protection at all times, not just when both lines are in service.

POWER RELAYS

At times the amount of power flow on lines must be held within a maximum limit, and protection is provided to cause the protected lines to trip by relay action in the event that power flow exceeds the predetermined limit. Relays sensitive to power flow are used for this purpose. Power relays are similar in construction to the directional element of directional overcurrent relays, except that the movable contact is restrained by a spring.

The torque developed by the current and voltage windings on the revolving disk is proportional to the power flow in the circuit being protected, and inverse-time characteristics (i.e., time for contact closure is reduced as power flow is increased) can be provided. Such relays can also be made directional, so that they will trip for power flow in one direction only.

DISTANCE RELAYS

A widely used and very successful method of protecting transmission lines is by so-called distance relaying. It is a characteristic of any electric circuit that the voltage drop in the circuit is proportional to the current in the circuit and the impedance of the circuit. Consequently, if a fault on a line occurs near a station bus, the current supplied to the fault will be greater than if the fault is farther from the station.

By making use of a voltage-energized coil to pull, on a balance beam, against the force produced by a coil energized by current proportional to line current, an impedance element is produced. When the impedance element is combined with a directional element, as has previously been described, a relay which is responsive to direction of power flow and distance (impedance) to a fault results. Other types of construction are used to obtain relays that combine the directional and distance features in one element.

In usual practice, distance relays are constructed to give three-zone protection. The first zone of station A includes approximately 90 percent of the section between two adjacent stations A and B. Zone 2 covers a portion (e.g., the last 10 percent) of the line between the stations protected by zone 1 and approximately 25 percent of the next line section. A third zone is provided, which covers the remainder of the line beyond the second-zone protection and reaches approximately 25 percent of the distance into the next line section to provide backup protection for the

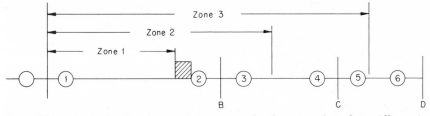

figure 10-6 Sketch showing the protection zones of a three-zone impedance (distance) relay. In the first zone, from breaker 1 to the crosshatched area, instantaneous operation will occur. In the second zone some time delay is introduced, and greater time delay occurs in the third zone. Similar relays would be installed at breaker 2 for protection toward breaker 1. The other line sections would be similarly protected.

first and second zones. The protection zones described above are indicated in Fig. 10-6.

One of the advantages of impedance relays is that they provide high-speed tripping operation by the circuit breakers at both ends of the line for most faults. When combined with other relays in a pilot scheme, they will provide nearly simultaneous operation at both line terminals. This is particularly important where high-speed reclosing is applied, as a time interval (approximately 12 to 20 cycles) with no power on the line is required so that the air at the point of the arc (fault) can deionize before the circuit is reenergized.

PILOT PROTECTION

High speed and additional selectivity can be added by providing carrier or pilot-wire schemes that will prevent tripping of breakers for faults outside the line section protected, but provide high-speed operation for all faults inside the line section. As mentioned previously, the first-zone protection of impedance relays provides simultaneous tripping from both ends of a faulted line section over approximately 90 percent of the line.

figure 10-7 With a fault in the line section between circuit breakers 3 and 4 as indicated, carrier signals would be sent from carrier transmitter 2 to receiver 1 and from carrier transmitter 5 to receiver 6. These breakers would be prevented from tripping, even though the direction is correct for tripping at 1 and 6 and the fault is within the distance protected by zone 2 of the relays for circuit breaker 6. The carrier signals are blocked between circuit breakers 3 and 4, as the fault is between them and near breaker 4. The zone 2 distance contacts at breaker 3 and the first-zone contacts of the relays at breaker 4 will provide instantaneous tripping of the faulted line section. Backup protection with time delay is provided by the zone 3 elements of the relays at circuit breakers 1 and 6.

With carrier-current blocking of impedance relays, the first-zone contacts operate independently of the carrier. The time-delay-element contacts are shorted out by carrier tripping contacts, permitting instantaneous operation for internal faults. The third-zone element operates independently of the carrier and provides time-delay backup protection and fault detection for control of a carrier or other pilot channel.

In the event of a fault in a line section, the carrier signal is started and immediately stopped by the directional or distance contacts; with no carrier signal received at the far end of the line, tripping of the circuit breaker is permitted. The action of carrier blocking is illustrated in Fig. 10-7.

PHASE-COMPARISON RELAYING

Another line-protection method that is used to provide high-speed operation is phase-comparison relaying. This system has some advantages over impedance (distance) relays, particularly where series capacitor compensation is used on long lines. With such compensation, the overall line impedance is less when series capacitors are in service than when they are out of service. Also, when a fault occurs near the capacitors, the impedance relays are not able to distinguish the location of the fault correctly. Series capacitor compensation is frequently used on long extra-high-voltage (375- to 750-kV) lines, and consequently a method of protection that will perform properly under all conditions is required for these installations.

In effect, a phase-comparison relay system compares current in and out of a line section. Under normal operation these currents must be equal. In the event of a fault on the line, the currents at both ends of the line are not necessarily equal, and the direction of power flow will reverse at the end of the line that had previously been receiving power.

Carrier pilot channels or microwave radio channels are used to transmit the relative phase positions of the currents at the two ends of the line, so that it can be determined whether the fault is internal or external to the protected line section. If the fault is internal, there will be an output from the electronic circuitry associated with the relay system, and tripping will result.

For external faults, the phase position of the received carrier signal at each end is such that the signals cancel with no output, so that the relays do not trip the line section.

DIRECTIONAL-COMPARISON RELAYING
SYSTEMS

Directional-comparison systems are frequently used for protection of transmission lines. These systems are also pilot-wire systems, and include both blocking and unblocking schemes, and overreaching and under-reaching transfer trip systems.

Directional-comparison blocking systems make use of directional distance phase and ground-fault detectors at each terminal. The fault detectors are set to reach beyond the terminals, and make use of pilot signals that are normally off. For a fault outside the line section, the fault detectors at the station adjacent to the fault will transmit a blocking signal to the remote station, where the direction would be correct for tripping, and prevent tripping. Tripping will not occur at the station adjacent to the fault, as the direction is wrong.

For internal faults, blocking signals will not be transmitted, and the circuit breakers at each end of the line can operate to clear the fault.

With these types of systems, blocking signals are transmitted only during fault conditions. Failure of the pilot signals will not prevent operation of the protection system, because the pilot signals are used to prevent tripping for external faults. However, as the system is overreaching, tripping for faults outside the protected section is possible after a time delay.

Directional-comparison unblocking systems transmit a continuous blocking signal except during internal faults. When internal faults occur, the blocking signal is removed (normally shifted in frequency), and the phase and ground-fault detectors will permit tripping to clear the faulted line section.

TRANSFER TRIP SYSTEMS

Transfer trip relaying is a method of protection whereby a tripping signal is transmitted to a remote line terminal, causing it to trip when a fault is detected in the protected line section.

Overreaching Transfer Trip Systems

Overreaching systems make use of a continuous pilot (guard) signal, and no tripping will occur while the guard tone is being received. A fault in the line section will cause the pilot frequency to shift to the trip frequency. At the same time the fault detectors at both ends of the line will operate, and trip signals will be transmitted to each line terminal, so that tripping will result to clear the line section.

Underreaching Transfer Trip Systems

Underreaching transfer trip systems may be either permissive or nonpermissive; they also make use of pilot signals. In nonpermissive systems the fault detectors are set to overlap in the protected line section, but not to respond to external faults. For internal faults, trip signals are transmitted from each end of the line to the opposite end, causing the circuit breakers to operate and clear the fault.

Permissive Transfer Trip Systems

In permissive transfer trip systems, in addition to fault detectors that reach only internal faults, additional overreaching phase and ground-fault detectors are used. For internal faults near one line terminal, the underreaching fault detectors at the remote terminal may not operate to send a tripping signal, but the overreaching detector will operate to send a tripping signal. At the station near the fault, the fault detectors will operate to send a tripping signal, so that trip signals are received at both ends of the line, and the circuit breakers will trip to clear the fault.

For external faults the fault detectors do not operate, and so no tripping will result.

The preceding discussion is by no means a complete description of power-line-protection methods, but is intended to outline briefly some of the principles commonly used for this purpose and some of the advantages and limitations associated with these various line-protection methods. More complete information is available in standard textbooks and manufacturers' literature pertaining to this subject.

DIFFERENTIAL PROTECTION

Before leaving the subject of power system protection, some mention should be made of some of the procedures used to protect station buses and equipment.

One of the most effective methods of protecting buses, transformer banks, and generating units is by differential relay circuits. Differential protection schemes are based on the principle that power into a bus or transformer bank, under normal conditions, is equal to power out. If the secondaries of current transformers are properly connected, under normal conditions, no current will flow in the relay current coil. Whenever a fault occurs, the current balance will no longer exist, and relay contacts will close, causing circuit breakers to operate to disconnect the equipment. The principle of differential protection for a transformer is illustrated in Fig. 10-8.

Station buses can be protected in the same manner. Current transformers of the same ratio are connected together so that the vector sum of all the currents will flow through the differential bus-protection relay. Under normal conditions, the power into and out of the bus will be equal, and the vector sum current will be zero. If a fault occurs in the

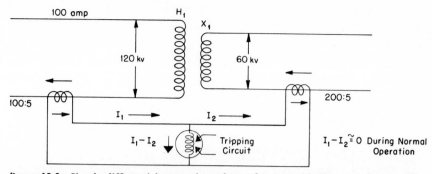

figure 10-8 Simple differential protection scheme for a two-winding transformer. The current transformer secondaries are connected in such a way that the current will circulate between them, and in normal operation no current will flow in the relay coil. In the event of internal trouble in the transformer, the current equality would no longer exist and current would flow through the relay coil, causing the contacts to close and trip the circuit breakers to deenergize the transformer.

protected area of the bus, power will flow into the fault and currents will no longer be equal, causing the relay contacts to close and trip all the breakers connected to the bus.

Generating units can be protected in the same manner by balancing the currents at both ends of each phase winding. Unbalance will result from a fault in the winding, causing the differential relays to operate and separate the machine from the system.

Many variations of differential protection have been developed for special applications, but basically they all operate on the principles just described.

ELEVATED NEUTRAL PROTECTION

Another method of protection frequently used with generating units is elevated neutral. This method of protection makes use of the fact that under normal operating conditions the phase potentials of a generator are fairly well balanced, and consequently the potential between the neutral lead and ground is near zero.

If unbalance should occur because of winding failure, the neutral voltage will increase. If the primary of a transformer is connected in series in the neutral lead, the voltage in the transformer secondary will be proportional to the neutral voltage. This voltage is used to cause a current to flow in a relay coil, and when the current exceeds a predetermined minimum the relay contacts will close and cause the generator circuit breaker to trip. This form of protection is illustrated in Fig. 10-9.

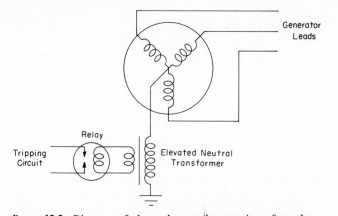

figure 10-9   Diagram of elevated neutral protection of an alter-
nating-current generator. Under normal (balanced) conditions, little or no current flows in the primary of the elevated neutral transformer. During unbalanced conditions, neutral current will flow,
and there will be an induced voltage in the secondary of the elevated
neutral transformer proportional to the voltage drop across the
impedance of the primary. The secondary voltage will cause current
to flow in the relay coil, and when it exceeds a predetermined
minimum, the relay contacts will close to trip the generator circuit
breaker.

INCORRECT OPERATION OF PROTECTIVE EQUIPMENT

An understanding of the principles of operation of protective devices
and systems should assist a system operator in determining whether or
not correct protective equipment action is obtained during system troubles. Such knowledge is also important in analyzing misoperations of
protective devices. A review and analysis of protective device operations
following a case of trouble can frequently assist in determining the cause
of the trouble.

Protective device operations should also be reviewed to determine
whether any devices operated that should not have done so or whether
any devices that should have operated failed to function.

As an example, in the legend accompanying Fig. 10-7 the correct
operations for the assumed trouble were outlined. It was pointed out
that breaker 6 would be prevented from operating as a result of the
carrier blocking signal from transmitter 5. However, if breaker 6 did
relay on the trouble in addition to breakers 3 and 4, it would be suspected
that either receiver 6 failed to receive the blocking signal or that transmitter 5 failed to transmit it. The incorrect operation of breaker 6 should,
of course, be investigated to determine the cause and correct it.

In the case of pilot protection, if it becomes necessary to remove the carrier equipment from one end of a line for any reason, such as bypassing a breaker, it is prudent to disable the carrier at the remote end.

When work is to be done on equipment protected by a differential relay installation, such as bypassing a breaker in a group protected by a bus differential setup, it may be necessary to cut out the differential protection to prevent incorrect relay operations.

Many other examples could be developed, and the importance of being familiar with proper protective device operation as well as the precautions to be taken to prevent erroneous operation should be apparent to all system operators.

SUMMARY

Various other protective methods are used on equipment, such as sudden pressure relays on transformers, overtemperature relays on transformers and generators, loss of field relays on generators, and the like. Further information on such protective devices can be obtained from manufacturers' literature.

A complete discussion of power system protection is far beyond the scope of this manual. The intent has been to provide some understanding of the basic principles of some of the more important and commonly used applications in order to provide power system operators with a knowledge of the type of protective equipment and action that they might expect from the protective installations on their systems. Such knowledge should make it possible to determine whether or not proper relay action occurs during trouble conditions.

PROBLEMS

1. A relay to be used across a line (i.e., one sensitive to voltage changes) will have a relay coil made up of
    (a) Few turns of a large conductor
    (b) Many turns of a small conductor
    (c) Few turns of a small conductor
2. Inverse-time relays
    (a) Are always provided with directional contacts
    (b) Require increased time for contact closure as current in the relay coil increases
    (c) Require less time for contact closure as current in the relay coil increases
    (d) Require less time for contacts to close as current in the relay coils decreases
3. When directional elements are included in a relay, the directional contacts will close when:
    (a) Current in the directional element coil exceeds a minimum value.
    (b) Power flow is in a predetermined direction.
    (c) Potential is applied to the directional element.

4. Ordinarily directional relays, for line protection, are connected so that the directional contacts will close for power flow
    (a) Away from a bus
    (b) Toward a bus
    (c) Exceeding a minimum amount
5. When balance relays are used for parallel-line protection and one of the parallel lines has a fault, the balance contacts will close to trip the faulted line
    (a) After a time delay
    (b) At once
    (c) When current in the faulted line exceeds a minimum
6. When distance relays are applied to protect a transmission line, relay protection is
    (a) Independent of the impedance to the fault
    (b) Confined to the first zone of protection
    (c) Normally provided in three zones
7. Impedance relays are combined with carrier or pilot-wire blocking in order to
    (a) Provide high-speed tripping for faults inside a protected line section
    (b) Provide carrier- or pilot-controlled time delays
    (c) Provide a signal that relay operation has occurred
8. One advantage of phase-comparison relaying is that:
    (a) Relay action can be provided for low values of fault current.
    (b) Correct relay action can be obtained with series capacitors in the line.
    (c) They will operate irrespective of the direction of power flow.
9. In transfer trip relay systems, the transmission of the transfer trip signal will cause
    (a) The circuit breaker at the station initiating the signal to trip
    (b) The circuit breaker at the remote station to trip
    (c) An alarm to be sent to the remote station
10. Differential protection of a generator makes use of the principle that under normal conditions:
    (a) The current at the neutral end of a phase winding is zero.
    (b) The currents in each of the phase windings are identical.
    (c) The currents at both ends of the phase winding are equal.
11. When an ac generator is protected by use on elevated neutral, in fault conditions of one phase:
    (a) The secondary voltage of the elevated neutral transformer becomes zero.
    (b) The secondary voltage of the elevated neutral transformer increases.
    (c) The phase potentials of the generator are balanced.

11

Power System Stability

INTRODUCTION

Stable operation of a power system requires a continuous match between energy input to the prime movers and the electrical load on the system. Load continuously changes as lights, appliances, or electrical machinery are connected and disconnected. Most individual load changes are small in relation to the size of the system, but each load increase or decrease must be accompanied by a corresponding change in input to the prime movers of the generators on the system.

If mechanical input does not rapidly match the electrical load, system speed (frequency) and voltage will deviate from normal.

More severe conditions exist when trouble occurs on a system, such as a fault on a transmission line or loss of a generator or a large block of load.

The function of the control equipment, such as governors, generator voltage regulators, and tie-line load-frequency-control equipment, is to sense deviations from normal and act to restore frequency and voltage to normal. Unfortunately, control devices are not perfect and usually permit oscillations (periodic variations) around the desired conditions. In most cases the oscillations diminish with time, and in such cases the system is stable.

Actually there are three stability conditions that must be considered:

1. Steady-state
2. Transient
3. Dynamic

To cover any of these topics fully is far beyond the scope of this manual. An effort will be made to provide a basis for an understanding of some of the factors involved in power system stability, so that system operators will have some appreciation of the subject. More complete information is available in standard textbooks and other technical literature.

STEADY-STATE STABILITY

Steady-state stability might be defined as the ability of an electric power system to maintain synchronism between machines within the system and on external tie lines following relatively slow or normally expected load changes. Steady-state stability to a great extent depends upon the margins of transmission and generating capacity and the effectiveness of the automatic control devices, particularly the generator automatic voltage regulators. The above statement is also true for transient and dynamic stability.

As described in the section "Transfer of Energy," whenever load on a synchronous generator is increased, the rotor will advance with respect to the revolving armature field, or drop back for load reductions. Normally the rotor angle change will slightly "overshoot," that is, retard or advance a little too much. Since it is not quite in the equilibrium position, it will continue, in a stable case, to oscillate in diminishing amounts until it settles down in the correct position for the new load condition. When the rotor settles to the correct position with only a few, rapidly diminishing oscillations, the machine is stable, and the oscillations are said to be highly "damped." Examples of stable and unstable conditions are shown in Fig. 11-1.

Swings of the type described are usually too fast for the machine governor to correct. However, fast-acting generator excitation systems (exciter and generator voltage regulators) can sense the voltage variations that accompany the rotor angle oscillations and strengthen or weaken the generator field in a time relationship to assist in achieving stable operation.

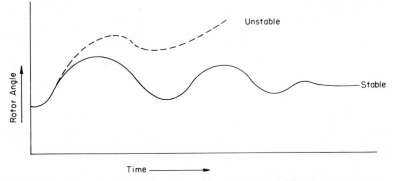

figure 11-1 Examples of stable and unstable swings of a synchronous generator following an increase of input to the prime mover and with constant field excitation.

The conditions just described are continuously present in a power system because many loads are being added or disconnected, and all the interconnected generators must continuously be readjusting energy inputs, rotor angles, and excitation to match the currently existing conditions.

TRANSIENT STABILITY

A more severe situation exists when a large block of generation or load is lost or a fault occurs on a transmission line. In such cases the transient stability of the system must be adequate to withstand the shock of the relatively large change that occurs. Transient stability is the ability of the system to remain in synchronism (prior to the action of governor control) following a system disturbance.

Immediately after loss of generation or load, the balance between energy input and electrical output of the system no longer exists. In the event of inadequate energy input, inertia of the rotors of the machines that remain in service will, for a short period, give up stored energy, and the machines will start to slow down. If load is lost, the energy input to the system exceeds the electrical load, and the machines will start to speed up.

Various factors affect the transient stability of a system, such as the strength of the transmission network within the system and of the tie lines to adjacent systems, the characteristics of the generating units, including the inertia of the rotating parts, and electrical properties such as transient reactances and magnetic saturation characteristics of the stator and rotor iron. Another important factor is the speed with which faulted lines or equipment can be disconnected and, with automatic

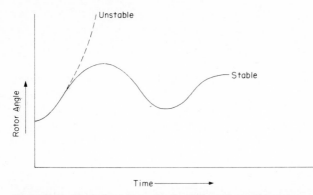

figure 11-2 Examples of transiently stable and unstable swings of a synchronous machine following a sudden disturbance. In the stable case the rotor angle increases to a maximum and then reduces. In the unstable case the rotor angle continues to increase until the machine exceeds the "pull out" position and loses synchronism.

reclosing of transmission lines, how rapidly lines can be restored to service. As in the case of steady-state stability, the speed with which the generator excitation systems respond is important in maintaining transient stability. System disturbances are usually accompanied by rapid reductions in system voltage, and rapid restoration of voltage to normal is important in maintaining stability.

As previously stated, transient stability is the ability to remain in synchronism during the period following a disturbance and prior to the time at which the governors can act. Ordinarily the first swing of machine rotors will take place within about 1 s following the disturbance, but the exact time depends on the characteristics of the machines and the transmission system. Following this period, governors begin to take effect, and dynamic stability conditions are effective. Examples of transient stability for stable and unstable situations are shown in Fig. 11-2.

DYNAMIC STABILITY

When sufficient time has elapsed after a disturbance, the governors of the prime movers will react to increase or decrease energy input, as may be required, to reestablish a balance between energy input and the existing electrical load. This usually occurs in about 1 to $1\frac{1}{2}$ s after the disturbance. The period between the time when the governors begin to react and the time when steady-state equilibrium is reestablished is the period during which dynamic stability characteristics of a system are effective. Dynamic stability is the ability of a power system to remain in synchronism after the "initial swing" (transient stability period) until the system has settled down to the new steady-state equilibrium condition.

During this period, the governors open or close valves, as required, to increase or decrease energy input to the prime movers, and the tie-line controllers operate to restore tie-line flows to normal. Usually, when the governors sense a speed drop, they will act to open the throttle valves to admit more steam to a steam turbine or water to a hydro unit and provide enough energy to arrest the decline of speed (frequency) and accelerate the system back to normal speed. This is still a condition of unbalance, because energy input now exceeds the load, and speed will increase to a point beyond normal, where the governors will again act to reduce energy input. As a result, oscillations of energy input and machine rotor angles will occur. If the system is dynamically stable, the oscillations will be damped, that is, reduced in magnitude, and after a few swings the system will settle down to an equilibrium (steady-state) condition with energy input equal to the electrical load on the system. This is illustrated in Fig. 11-3.

It is possible to have transiently stable but dynamically unstable conditions. Immediately after a disturbance, the machine rotors will go through the first swing (before governor action) successfully; then, after governor control is initiated, the oscillations will start to increase until the machine goes out of synchronism. This can occur if the time delays

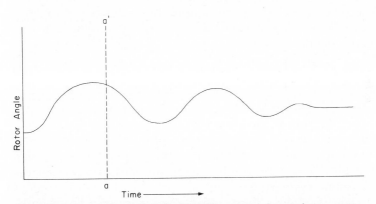

figure 11-3 Example of dynamically stable swings of a synchronous machine following a disturbance. During the first swing, from the time of the occurrence of the disturbance to the time indicated by the dotted line a–a', the governors have not had time to react. This is the period during which transient stability is important. After governor action occurs, the machine rotor angle will oscillate with diminishing swings until it settles down at the new angle corresponding to the new load requirements. This time may extend from about 30 s to several minutes, depending on the inertias of the system and the characteristics of the governors.

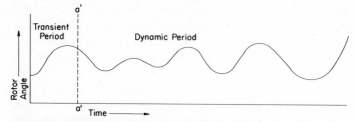

figure 11-4 Example of a transiently stable but dynamically unstable condition. During the transient period (the first swing) the machine angle reached a maximum and recovered. During the period of governor control action the oscillations again increased with time, and if they continue the machine will go out of synchronism.

of the governor control are such that, following the sensing of need for increasing or decreasing energy input, action is delayed sufficiently in time to augment rather than diminish the next swing. If such a condition exists, the oscillations of the machine rotor can continue to build up until the machine drops out of synchronism. Such a condition is illustrated in Fig. 11-4.

Investigations of stability at one time included only the first (transient) swing, and it was thought that if this was stable the oscillations would damp out. Until the advent of digital computers, with their capability of high-speed calculations, stability studies were very time-consuming. At present it is common to carry the studies through several swings to ensure that the system is stable and that the swings will diminish with time to a steady-state condition.

SUBSYNCHRONOUS RESONANCE (SSR)

A special condition can sometimes exist on long lines, usually of ehv, with no intervening taps or connections to the system between the generating station and the receiving substation. Such long lines may have a resonant frequency somewhat below the system operating frequency. Large thermal machines may also have resonant frequencies below the power system operating frequency.

Under certain conditions the resonant frequency of a line may be approximately equal to that of the machines. In such an eventuality, changes in the rotor angles of the machines may be at a frequency at or near the resonant frequency of the line, and oscillations of increasing magnitude can occur. This condition is known as subsynchronous resonance. In this condition interaction between the line and the generators can cause increasing power swings, and, if sustained, these can cause damage to the generators connected to the line. There have been instances in which machine swings have increased to such an extent that machine shafts have broken.

Present practice in system design includes studies and tests to determine whether or not SSR conditions may exist, and corrective measures are taken to eliminate or minimize the possibility of problems from SSR.

Corrective measures include the addition of SSR filters, which are inserted in the lines when loading is such that SSR oscillations could occur. The filters are inserted in steps, depending on line loading. At light loads there is little danger of subsynchronous oscillations.

SUMMARY

The system operator cannot do much about stability, since field-voltage regulators, governor systems, and the transmission network are part of the system design. However, in day-to-day operation the operator can avoid some conditions that might lead to unstable operation, such as operating machines with weak fields or loading transmission or tie lines near their stable limits so that there is insufficient margin to ride successfully through transient disturbances.

PROBLEMS

1. If mechanical input to prime movers of generators of a power system does not match load changes:
   (a) System losses will be increased.
   (b) System frequency will be low.
   (c) System frequency and voltage will deviate from normal.
2. Steady-state stability of a power system is the ability of a system to
   (a) Maintain frequency at exactly 60 Hz
   (b) Maintain synchronism between machines and on external tie lines
   (c) Maintain a spinning-reserve margin at all times
3. When a synchronous generator is operating stably and a sudden load change occurs, the rotor will
   (a) Advance in its position relative to the stator revolving field
   (b) Not be affected
   (c) Oscillate in diminishing amounts until it settles to its new correct position
4. When the rotor of a synchronous generator oscillates because of load changes, corrective action can be provided by
   (a) Fast-acting generator voltage regulators
   (b) The governor system
   (c) Manually reducing the field current
5. Transient stability of a power system is the ability of the system to
   (a) Deliver power over tie lines to other systems during a disturbance
   (b) Remain in synchronism, prior to governor action, following a system disturbance
   (c) Maintain a continuous balance between energy input and electrical output
6. Transient stability of a power system is affected by
   (a) The characteristics of the generating units
   (b) The strength of the transmission network

    (c) The speed with which faulted lines can be disconnected

    (d) All of the above

7. Dynamic stability characteristics of a power system are effective in the period

    (a) Immediately following a loss of load or generation

    (b) While governor action is taking place

    (c) Between the time at which governors begin to react and the time at which steady-state equilibrium is established

8. When a system is dynamically stable, oscillations of synchronous machines following sudden load changes will

    (a) Diminish with time

    (b) Increase with time

    (c) Not occur

9. If a power system is transiently stable, it

    (a) Will always be dynamically stable

    (b) May be dynamically unstable

    (c) Will never oscillate beyond the first swing

10. When a power system is dynamically unstable:

    (a) Oscillations may increase until generating units go out of step or tie lines relay.

    (b) Governor action has no effect.

    (c) Energy input to the system always exceeds the electrical load of the system.

11. Subsynchronous resonance is a condition that may occur on

    (a) Short transmission lines

    (b) Long, heavily loaded lines

    (c) Loss of load on a transmission line

12

EHV Operation

INTRODUCTION

In recent years there has been a great deal of activity in the power industry in construction of extra-high-voltage (ehv) lines and stations. Ehv is usually considered to be lines and equipment operating above 230 kV. A considerable amount of 345-kV transmission is in existence, and more recently 500-kV and 750-kV transmission lines have been built. Research is in progress for transmission of power at voltages up to 1000 kV.

As the amounts of power to be transmitted increase and as the distances between points of generation and load become greater, it becomes necessary to increase transmission voltages. Factors limiting transmission voltage are the availability of transformation equipment, line insulation, and switching devices capable of operating at extremely high voltages.

Beginning in the early 1920s and continuing through the early 1960s, 230 kV was the usual maximum transmission voltage on large power systems and for major interconnections between power systems. With

230-kV lines of a hundred or so miles in length, the maximum load capability is usually of the order of 200 or 300 MW. Shorter lines of this voltage may carry up to 400 MW or more.

With the availability of larger generating units of 500- to 1000-MW capacity and with increased sizes of power systems, it became desirable to transmit much greater amounts of power on electrical transmission lines beyond the capability of the usual 230-kV transmission systems.

When it became apparent that higher transmission voltages would be required, electrical equipment manufacturers and power utilities initiated research projects to develop switch gear, transformation equipment, and methods of insulating and constructing lines to operate in the 500- to 750-kV range.

Basically, ehv lines and equipment follow the same electrical laws that are effective at lower voltages. The ehv circuits still contain only electrical resistance, inductance, and capacitance, which are present in all electric circuits.

SPECIAL REQUIREMENTS FOR EHV

Problems that required solution before ehv transmission became practicable were:

1. The limitation of corona losses and techniques for maintaining radio interference at tolerable values
2. Development of power circuit breakers capable of interrupting extremely high currents at the operating voltage levels
3. Effective methods for protecting lines and equipment
4. Development of adequate disconnecting switches and line and station hardware

CORONA DISCHARGES

When a conductor is charged to an electrical potential (voltage), the insulating medium surrounding the conductor (air in the case of overhead lines) is stressed electrically. The electric field surrounding a conductor can be measured at various distances from the conductor. When such measurements are made, it is found that the voltage between equidistant points in space, at a right angle to the direction of the conductor, decreases as the distance to the conductor increases. The electric charge in space between two points is called the potential gradient.

Although air is a nearly perfect insulator at low voltages, as the potential gradient is increased, the air molecules become ionized and conductive. When the ionizing potential is exceeded, current that would normally flow in the line is dissipated in the atmosphere, resulting in energy losses. This loss is called corona and, when sufficiently great, is visible as a glow around the conductors. In addition to the power loss resulting from corona, there are high-frequency (radio-frequency) components associated with the corona discharge which cause radio interference.

Several factors affect the voltage level at which corona discharges will occur. As the conductor diameter is increased, the potential gradient around the conductor is decreased, making it possible to operate a line without corona at a higher voltage than would be possible with a small-diameter conductor.

Another factor affecting corona is the smoothness of the conductor. Any roughness on the conductor will produce points of high potential gradient, with attendant increase in corona loss. This is also true if there are sharp points or corners on hardware assemblies.

BUNDLED CONDUCTORS

As mentioned previously, one method of reducing potential gradient around a conductor is to increase the conductor diameter. At 230 kV conductors of approximately 1 in diameter provide a satisfactorily low corona-loss level. At 500 kV or higher, conductors several inches in diameter would be required to obtain a reasonably low corona loss. Conductors of such size would be very expensive, but an even greater problem would be the size of the physical structures needed to support the weight of such conductors. Another problem would be the development of suspension insulators capable of supporting conductors of the size and weight required.

It has been found that when two or more reasonably sized conductors are spaced a few inches apart and operated in parallel, the "bundled" conductors behave as one large conductor, as far as the potential gradient in the surrounding atmosphere is concerned. The potential gradient around each of the individual paralleled conductors merges with the gradients of the other conductors so that the composite gradient (electric field) is similar to that which would exist with one extremely large-diameter conductor.

By bundling conductors, it became possible to increase operating voltages up to the 500- to 750-kV range economically and with tolerable corona losses and radio influence.

POWER CIRCUIT BREAKERS

For many years, automatic circuit breakers used in power-transmission systems were oil-filled. At so-called ehv levels, oil circuit breakers became impractical, and interruptors were developed that operate at line potential on insulated columns. In order to interrupt voltages of 500 kV and higher, several interruptors are usually connected in series. For quenching the arcs resulting when a power circuit is interrupted in these applications, high-pressure dry air or a gas such as sulfur hexafluoride (SF_6) is used in place of insulating oil.

EFFECTS OF LINE REACTANCE ON LINE-VOLTAGE REGULATION

In Chap. 2 it was pointed out that an electrical transmission line, in addition to conductor resistance, always contains series inductance and shunt capacitance. As line loading increases, the voltage drop due to current flowing through the series inductive reactance $E_{X_L} = IX_L$ becomes appreciable, resulting in a receiving-end voltage considerably below the sending-end voltage. Also, because current through an inductance is delayed (see Fig. 1-1), the phase angle between the sending and receiving ends of the line may reach the limit of stability considerably before the load-carrying capability of the line is reached. Figure 12-1 shows an approximate vector diagram of a transmission line where current lags the voltage. This diagram is approximate, since the inductance and capacitance of a transmission line are distributed throughout its length, and an exact representation of a long line is considerably more complex than indicated in Fig. 12-1. However, this figure should serve to indicate the effect of inductive reactance of a line.

As has previously been noted, capacitance exists between conductors

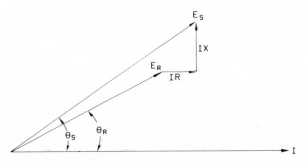

figure 12-1 Approximate vector diagram of a single-phase transmission line operating at a lagging power factor. The receiving-end voltage E_R is less than the sending-end voltage E_S by the vector sum of the voltage drops in the line caused by current through the line resistance and the line reactance. Θ_S is the angle between the current and voltage at the sending end, and Θ_R is the angle between the current and voltage at the receiving end of the line. The difference between Θ_S and Θ_R is the phase shift in the line.

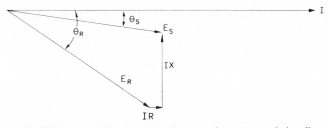

figure 12-2 Approximate vector diagram for a transmission line operating with a leading power factor. The vector sum of the voltage drops in the line results in a receiving-end voltage greater than the sending-end voltage.

of a line and between the conductors and ground. When a line is lightly loaded, the capacitive charging current may exceed the load current and result in the line's operating with a leading power factor. In such cases, the receiving-end voltage will rise, and may exceed the voltage at the sending end of the line. This condition is indicated in the approximate vector diagram shown in Fig. 12-2.

In lines with bundled conductors, the capacitances involved are greater than those in lines with single conductors, resulting in increased capacitive charging currents. Furthermore, as line voltage is increased, the capacitive charging kvar of a line increases. For example, on a 500-kV line with two bundled conductors, approximately 2000 kvar/mi of capacitive reactive supply is required.

If a long line is lightly loaded, its receiving-end voltage will rise above that of the sending end. If it is heavily loaded, the receiving-end voltage will drop considerably below the sending-end voltage. When the voltage rise at light loads is excessive, insulation may be overstressed or voltage-regulating equipment at the receiving-end station may go out of range, resulting in undesirable customer voltage conditions.

When a long line is heavily loaded and its receiving-end voltage drops excessively, voltage-regulating equipment may go out of range and cause customer voltage to be reduced.

REACTANCE COMPENSATION WITH SERIES CAPACITORS

In order to minimize the adverse conditions mentioned above, series capacitors and shunt reactors are frequently used on long ehv lines.

When series capacitors are installed in a line, the voltage drop through the series capacitive reactance is proportional to the line current, and the capacitive reactive drop is vectorially opposite to the drop through the line inductive reactance. The result is reduced voltage regulation at the receiving end of the line. This is shown in the vector diagram in Fig. 12-3.

From Fig. 12-3 it can be seen that the receiving-end voltage with series capacitor compensation E_{R_1} is considerably higher than the voltage that would exist without the use of series capacitors, E_{R_2}. It should also be noted that the phase shift between the sending and receiving ends of the line is reduced when series capacitors are used. In effect, the series capacitors shorten the line electrically and make it possible to carry more load without instability than would be the case without them.

Series capacitors as used for high-voltage line compensation are made up of groups of capacitor units similar to those used for power-factor correction on distribution lines and connected in series-parallel. The number of units connected in series is sufficient to withstand the maximum voltage drop expected across the capacitors. The number connected in parallel is determined by the normal line currents to be expected. The series capacitors must be operated at line voltage, and they are mounted on platforms with sufficient insulation to withstand the line-to-ground voltage of the transmission line.

Series capacitor installations can be made with all the capacitors at one location, such as the midpoint of the line section, or with one-half the capacitors at each end of the line. Both types of installations are in service, and the decision where to install capacitors is made on the basis of engineering and economic considerations. Figure 12-4 shows a single-line schematic diagram of both types of installations.

Actual installation of series capacitors is considerably more complicated than Fig. 12-4 indicates. Disconnecting, bypass, and ground devices

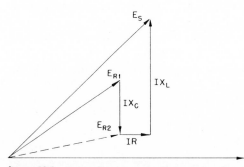

figure 12-3 Approximate vector diagram showing the effect of the use of series capacitor compensation of a transmission line. The vector E_{R1} represents the receiving-end voltage with series capacitor compensation in service, and the dashed line E_{R2} represents the receiving-end voltage without series capacitor compensation.

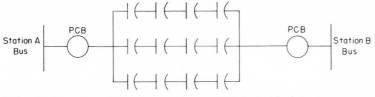

figure 12-4(a)   Line with series capacitors installed at midpoint.

figure 12-4(b)   Line with one-half the series capacitors installed at each end.

are included at each installation, and gaps are provided across the capacitors that will break down if the voltage drop exceeds a predetermined maximum, which might occur under fault conditions.

REACTANCE COMPENSATION WITH SHUNT REACTORS

As was pointed out earlier in this section, under light load conditions the capacitive charging current may cause excessive voltages at the receiving end of a long high-voltage line. This situation is the exact opposite of that which occurs under heavy load conditions where series capacitors are used for compensation. It should be recalled from the discussion in Chap. 1 that inductive reactance is opposite in effect to capacitive reactance.

In order to compensate for the voltage rise resulting from capacitive charging currents, inductive reactances can be installed from line to ground or across the tertiary windings of transformer banks. Such installations draw lagging current and correct for the voltage rises that occur in the lines under lightly loaded conditions.

A shunt inductive reactor for line-to-ground installation is similar to the primary winding of a high-voltage step-down transformer, but no secondary windings are included. Such reactors are wound on an iron core and installed in oil-filled tanks. The winding must have a sufficient number of turns to provide the desired inductive reactance.

When shunt reactors are installed on transformer bank tertiaries, they are usually air-core coils with enough turns to provide the desired reactance for the operating voltage. Figure 12-5 shows single-line schematic diagrams of both types of reactor installations.

As indicated in Fig. 12-5, shunt reactors are provided with switching

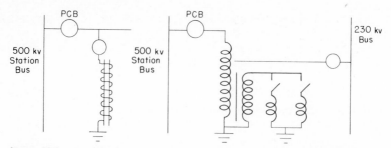

figure 12-5   One-line schematic diagrams of shunt reactor installations. (*a*) A full-voltage reactor with an iron core. (*b*) Reactors of the tertiary winding of a transformer bank.

devices so that they can be connected or disconnected as required to compensate for undesirable voltage conditions. Typically, high-voltage reactors (for example, 500 kV) might be 35 to 40 Mvar each. Tertiary reactors are limited in size by the amount of inductive current that can be switched at the tertiary voltage, and are ordinarily approximately 15 Mvar each.

PROTECTION OF LINE WITH SERIES CAPACITORS

One other factor that should be mentioned in connection with ehv lines is that when series capacitor compensation is used, normal distance (impedance) relays, like those discussed in Chap. 10, cannot be used effectively. Under fault conditions the relays may not be able to distinguish the location of the fault and may fail to trip, causing a delay in deenergizing the circuit. Phase-comparison relays, also discussed in Chap. 10, are frequently used for ehv-line protection, since they are capable of providing selective high-speed protection which both clears faults rapidly and helps maintain the stability of the power system or of interconnections to other systems.

SUMMARY

The preceding discussion on ac ehv lines may be summarized as follows:

1. Operation at extra-high voltages still follows basic electrical laws.

2. At these voltages special techniques are required to reduce potential gradients in order to minimize corona losses and radio influence.

3. Because of greater line lengths, loads, and charging currents, it is frequently necessary to use series capacitor or shunt reactor compen-

sation to maintain line stability and receiving-end voltage within desired limits.

4. The use of series capacitors on ehv lines requires special protection consideration as compared with their use on more conventional lower-voltage lines.

EHV DC TRANSMISSION

This discussion so far has been concerned with ac systems. Installations have also been made and others are planned for dc transmission lines operating in the 500- to 1000-kV range. The first transmission of electric power was by direct current. In the early days of the power industry there was considerable controversy over the merits of dc versus ac systems. With the development of the transformer, it became possible to raise ac voltages to such values that relatively large quantities of power could be transmitted over considerable distances economically. As a result, almost all high-voltage transmission has been by alternating current. The notable exception was in Europe, where some installations of high-voltage (approximately 100-kV) direct current were made by the Thury system. In this case dc generators were connected in series, and the machines, mounted on insulated platforms, operated at or near line potential.

RECTIFICATION AND INVERSION

The development of grid-controlled mercury-arc rectifiers and silicon controlled rectifiers (SCRs) has made it possible to produce direct current at high power levels with high efficiency and without the problems of mechanical commutators required in rotating machinery.

High-voltage dc transmission is accomplished by operating grid-controlled mercury-arc rectifiers or SCRs in series-parallel to develop the desired voltage levels and required current-carrying capabilities.

Grid-controlled rectifiers and SCRs can be used for rectification of alternating current to direct current. Developments in solid-state technology have progressed to the point that grid-controlled mercury-arc rectifiers have been supplanted by SCRs in modern ehv dc installations. By triggering these devices in proper time sequence, they can be used as inverters to convert direct current back to alternating current.

Figure 12-6 shows a simplified schematic diagram of a rectifier-inverter using SCRs.

The SCRs shown in Fig. 12-6 are numbered in their normal firing sequence. When the grid-control pulses are timed so that the SCR fires

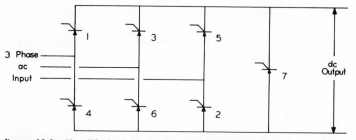

figure 12-6 Simplified schematic diagram of Graetz bridge circuit used on SCR in rectifier-inverter installations for high-voltage dc transmission systems.

(conducts) in the first quadrant of the ac cycle, it will act as a rectifier. If the control pulse is timed so that the SCR fires in the second quadrant of the ac cycle, the SCR will function as an inverter. This timing function is shown in Fig. 12-7.

Since power generation and distribution are almost universally by alternating current, the use of dc transmission lines is limited to transporting large blocks of power from one location to another within a system or between systems. In all applications of high-voltage dc transmission, ac voltage is stepped up by conventional transformers before rectification. It is then transmitted as direct current to the point at which it is inverted again to alternating current and passed through transformers to obtain the desired ac voltage level.

It should also be pointed out that if the rectifiers and inverters are identical, current is rectified or inverted as desired by timing pulses on

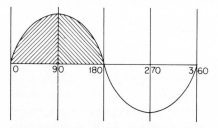

figure 12-7 Representation of a control cycle for a rectifier-inverter installation. When the firing pulse is initiated in the first quadrant, 0 to 90°, the valve will act as a rectifier. When the pulse is initiated in the second quadrant, 90 to 180°, it will act as an inverter, providing an ac output from a dc input. The amount of power flowing through the rectifier or inverter is controlled by the timing of the firing pulse.

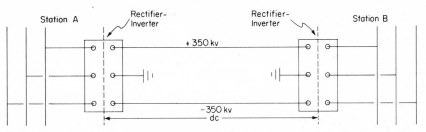

figure 12-8 Simplified schematic diagram of a 700-kV dc transmission line connecting two ac stations. For power flow from station *A* to station *B*, the rectifier-inverter equipment at station *A* would rectify to produce direct current, and the equipment at station *B* would invert to convert direct current back to alternating current. For power flow from station *B* to station *A*, the operation of the rectifier-inverter equipment would be reversed.

the SCRs. Figure 12-8 shows a simplified schematic diagram of a dc transmission line between two ac systems.

GROUNDED OPERATION

Most high-voltage dc transmission lines operate grounded. In this arrangement, one conductor operates at a positive potential (voltage) equal to one-half the line-to-line voltage, and the other conductor operates at a negative potential equal to one-half the line-to-line voltage. Under normal balanced conditions, little or no current flows through the ground.In the event of trouble, such as flashed insulators or a damaged conductor, however, the line can be operated at reduced capacity with only the positive or negative conductor in service and with the ground serving to complete the circuit.

CORROSION PROBLEMS

When a dc line is operated with one conductor against ground, the current through the ground may cause severe corrosion of buried gas pipelines, water systems, communication cables, and other underground installations. This problem is similar to those that exist in cities with street railway systems, which usually operate at approximately 600 V dc from an insulated trolley wire. Current return is through tracks and the earth. Various measures are taken to eliminate or minimize corrosion resulting from the railway operation. Such measures include the use of rectifiers connected to the buried pipes to charge the pipes at a potential equal and opposite to that due to the earth currents from the railway system. High-voltage dc lines may present more severe problems because of the magnitudes of the earth currents involved.

ADVANTAGES OF DC TRANSMISSION

Several factors favor high-voltage direct current for transmission of large blocks of power over long distances. Ac lines are rated at what is called the "effective voltage." The peak voltage of the wave is approximately 1.4 times the effective voltage, so that a line rated at 500 kV actually has to be insulated to stand 1.4 × 500 or 700 kV line to line. The maximum line-to-ground voltage is approximately 405 kV. A 700-kV dc line operating grounded as discussed above would be 350 kV positive and 350 kV negative with respect to ground. As a result, a lower insulation level is required.

In dc transmission there are no capacitive- and inductive-reactance effects, and the loss in the line is all due to its resistance.

Another advantage of dc transmission is that paralleled systems do not have to operate in synchronism, as is necessary on ac systems. Power transfer is adjusted by increasing the voltage at the sending end as described in Chap. 2.

Because dc systems operating in parallel do not have to operate in synchronism, dc interconnection can be used to transfer power between systems of different frequency. This has been done in Japan, where a 60-Hz system operates in parallel with a 50-Hz system via a dc interconnection. In North America, however, power transfer between systems is not normally a problem, because almost all systems now operate at 60 Hz.

DISADVANTAGES OF DC TRANSMISSION

A principal disadvantage of dc transmission is the cost and complexity of the rectifier-inversion equipment. This factor makes it expensive to install intermediate substations on a dc line, whereas in ac transmission it is relatively inexpensive to install transformer banks at intermediate stations to supply load requirements.

The choice between ac and dc transmission is determined by economic and engineering factors to best meet the requirements of a particular installation.

PARALLEL OPERATION OF AC AND DC SYSTEMS

Ac and dc transmission lines may operate in parallel when proper equipment is provided to control the loading and division of load between the lines. Parallel operation of 500-kV ac and 750-kV dc lines has been accomplished with the Pacific Intertie facilities between the states of Oregon and California, and by various other installations in the United States and other countries.

Several advantages result from parallel operation of ac and dc lines:

1. Power flow over the dc line can be set and remain independent, within reasonable limits, of the power angle or system voltages.

2. The desired power can be delivered by the dc line to an ac bus without appreciably increasing the short-circuit duty on the bus.

3. With a bipolar dc line, that is, a line operated grounded and with one conductor positive and the other conductor negative with respect to the ground, the line reliability is equivalent to that of a double-circuit ac line. This is true because either pole can continue to operate when the other pole is out of service because of line or terminal filter outages.

Almost all dc line installations are two-terminal lines. Conventional ac circuit breakers are used on the ac sides of the rectifier-inverter equipment. One reason multiterminal dc lines are not normally installed is that dc high-voltage circuit breakers are not available. High-voltage dc transmission probably will be limited to two-terminal lines until satisfactory dc circuit breakers are developed. However, work is being done to develop multiterminal dc lines.

SUMMARY

In summary, high-voltage dc transmission presents advantages in some cases, and may be used to transmit large blocks of power over long distances. Parallel operation of ac and dc systems can be accomplished when proper control devices are provided.

PROBLEMS

1. Ehv lines and equipment
    (a) Do not follow the same electrical laws that are effective at lower voltages
    (b) Contain only inductance and capacitance
    (c) Follow the same electrical laws that are effective at lower voltages
2. When electrical potential of a conductor is increased:
    (a) The potential gradient around the conductor increases.
    (b) The potential gradient around the conductor decreases.
    (c) Corona losses are reduced.
3. Corona losses are minimized when:
    (a) Conductor size is reduced.
    (b) Sharp points are provided on the line hardware.
    (c) Smooth conductor is used.
4. When two conductors are "bundled":
    (a) The potential gradient in space is increased.
    (b) Corona losses are eliminated.
    (c) The potential gradient around them is equivalent to that which would exist with one large-diameter conductor.
5. When a long transmission line is heavily loaded:
    (a) The distributed capacitive reactance of the line predominates.
    (b) The inductive reactance and resistance of the line cause the receiving-end voltage to be less than the sending-end voltage.
    (c) Corona losses are increased.

6. On a lightly loaded long transmission line:
   (a) The receiving-end voltage may exceed the sending-end voltage.
   (b) The receiving-end voltage always exceeds the sending-end voltage.
   (c) The capacitive charging current is reduced.
7. When series capacitors are used in a transmission line:
   (a) The phase angle between the sending and receiving ends is increased under heavy load conditions.
   (b) The phase angle between the sending and receiving ends is reduced under heavy load conditions.
   (c) Line stability is impaired.
8. The voltage drop across a series capacitor bank is proportional to
   (a) The voltage of the line
   (b) The size of the line conductor
   (c) The current flowing in the line
9. Shunt reactors are installed at the terminals of long high-voltage transmission lines in order to
   (a) Increase the terminal voltage under heavy load conditions
   (b) Compensate for voltage rises caused by capacitive charging currents at light loads
   (c) Compensate for the power factor of the connected loads
10. When ehv dc transmission is used:
    (a) The power is generated by dc generators.
    (b) Ac power is rectified at the transmission voltage.
    (c) Rotating machines with mechanical commutators are required.
11. At the terminals of ehv dc transmission lines, rectification or inversion is accomplished by
    (a) Grid-controlled mercury-arc or silicon controlled rectifiers
    (b) Use of series capacitors
    (c) Motor-generator installations
12. When a dc transmission system is grounded, with one conductor positive with respect to ground and the other conductor negative, the amount of current flowing through the ground is
    (a) Proportional to the voltage of the line
    (b) Proportional to the line current
    (c) Proportional to the unbalance of the currents in the two conductors
13. On a dc transmission line the maximum line-to-ground voltage is
    (a) Equal to the peak voltage of an ac transmission line of the same nominal voltage rating
    (b) Less than the peak voltage of an ac transmission line of the same nominal voltage rating
    (c) Greater than the peak voltage of an ac transmission line of the same nominal voltage rating
14. In a dc transmission line:
    (a) The effects of inductive and capacitive reactance are greater than in an ac transmission line of the same rating.
    (b) It is necessary for the sending and receiving ends to be operated in synchronism.
    (c) There are no effects due to capacitive and inductive reactance.
15. The choice between ac and dc transmission is determined by
    (a) Engineering and economic factors
    (b) The size of the conductors
    (c) The operating voltage

An Introduction to Trigonometry

INTRODUCTION

Trigonometry, which is the mathematics of angles and triangles, is used a great deal in ac circuit analysis. A complete study of the subject can get quite involved, but for most problems in day-to-day power system operation, basic and very simple applications of trigonometry can be of major assistance.

The following paragraphs will develop the subject sufficiently that trigonometric principles can be applied to the solution of simple electrical problems.

TRIGONOMETRIC FUNCTIONS

Four ratios can be developed and tabulated that make the solution of triangles very simple. Strange as it may seem, triangles frequently occur in problems involving ac currents, voltages, power, and reactive (var) flows.

If a circle with a radius of 1 unit (inch, foot, mile, etc.) is drawn and a right triangle is formed as in Fig. A-1-1, the basis for two important trigonometric functions (properties) can readily be developed. The side *ac* is the hypotenuse. The other two sides are the projection *ab* on the horizontal radius and the side *bc* connecting the projection and the point of intersection of the hypotenuse.

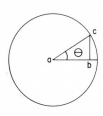

figure A-1-1 Circle with a radius of 1 and right triangle *abc* inscribed in it. The angle between the sides *ab* and *ac* is denoted with the Greek letter theta (Θ). By definition, the length of the side of the triangle away from the center of the circle *bc* divided by the length of the hypotenuse *ac* is called the sine (abbreviated sin) of the angle, or

$$\sin \Theta = \frac{bc}{ac}$$

Also, the side adjacent to the center *ab* divided by the hypotenuse *ac* is called the cosine (abbreviated cos), or

$$\cos \Theta = \frac{ab}{ac}$$

If a circle is drawn to a reasonably large scale on graph paper with the aid of a protractor, it is very simple to determine the sines and cosines for various angles. In order to make going through this procedure unnecessary, tables have been developed showing these quantities. By inspection, a few basic properties of sines and cosines can be determined. If Θ is 0°, the side of the triangle *bc* is 0 and the sine of Θ is 0. Also at 0°, the side *ab* is 1 and the cosine of 0 is 1. At 45°, the sides *bc* and *ab* are equal and if measured would be 0.707 unit long, so that at 45°, sin Θ and cos Θ are equal and are 0.707. At 90°, the side *ab* is 0 and the side *bc* is 1, so that at 90°, sin Θ = 1 and cos Θ = 0. At 135° (90 + 45), sin Θ = 0.707 and cos Θ = 0.707. At 180°, sin Θ = 0 and cos Θ = −1. At 225°, sin Θ = −0.707 and cos Θ = −0.707. At 270°, sin Θ = −1 and cos Θ = 0. At 315°, sin Θ = −0.707 and cos Θ = 0.707.

It can readily be seen that sines and cosines repeat in value, but are positive or negative depending on what quadrant of the circle they are located in. Figure A-1-2 shows this relationship.

Very shortly it will be shown how sines and cosines can be used in electrical problems, but before proceeding with the application of these quantities it is desirable to develop the other two basic trigonometric

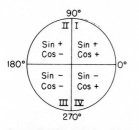

figure A-1-2 Relationships of algebraic signs of sines and cosines in various quadrants of a circle.

ratios. Again we will draw a circle with a 1-unit radius and draw a triangle so that the side opposite is at the end and perpendicular to the horizontal radius, as shown in Fig. A-1-3.

If Θ is 0°, the side bc is zero and the ratio bc/ab is 0, or at 0°, tan Θ = 0. However, the ratio ab/bc is infinity ($1/0 = \infty$), and cot $\Theta = \infty$. At 45°, ab and bc are equal, so that tan $\Theta = 1$ and cot $\Theta = 1$. At 90°, the side bc would be infinitely long, and the side ab would still be 1, so that the ratio bc/ab would be $\infty/1$ or tan 90 = ∞. The values of the tangents and cotangents would repeat in the other quadrants except for their algebraic signs. The signs are shown in Fig. A-1-4.

It is probably obvious by now that since the tangent is bc/ab and the cotangent is ab/bc, if the tangent is known, the cotangent can be obtained by dividing the tangent into 1, or cot = 1/tan.

Now that the four important trigonometric ratios have been defined, we will apply them to a few everyday electrical problems. It should be emphasized that the ratios just defined hold for any size circle or triangle. Therefore, if the sines, cosines, tangents, or cotangents are known, they can be used as multipliers or divided into sides of any right triangle, as applicable, to find unknown sides; conversely, given the sides of a right triangle, any of the ratios can be determined. So that it will not be necessary to determine the ratios from measurements on a unit circle

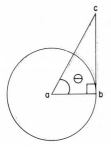

figure A-1-3 Circle with radius of 1 and triangle with the side opposite at a right angle to ab and touching the circumference of the circle. The ratio of the length of the side bc to the side ab, bc/ab, is the tangent (tan) of the angle Θ. Also, the ratio of the side ab to the side bc is the cotangent (cot) of Θ.

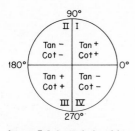

figure A-1-4 Relationships of algebraic signs of tangents and cotangents in various quadrants of a circle.

with a protractor and graph paper, a brief table of trigonometric functions by degrees is given in Table A-1-1. To use this table for angles between 0 and 45°, read from the top down, and for angles from 45 to 90°, read from the bottom up. The column headings at the top and bottom of the table show which function is in each column.

Tables are available giving the functions to fractions of a degree (minutes or seconds) or by decimal parts to tenths or hundredths. For purposes of this manual Table A-1-1 is adequate.

Electronic calculators of the so-called scientific type provide trigonometric functions, making it unnecessary to refer to tables. Such calculators make the solution of problems involving trigonometry very simple.

APPLICATION OF TRIGONOMETRIC FUNCTIONS

Now to solve a few everyday problems:

1. The load on a circuit is 1500 kVA, and the power factor is 78 percent lagging. What is the kilowatt load, and how many kilovars are being supplied to the bus? Power factor is actually the cosine of the angle between the apparent power (kVA) and actual power (kW). From what we know about trigonometry the problem is simple.

$$(a)\ \mathrm{kW} = \mathrm{kVA} \times \cos \Theta$$
$$= \mathrm{kVA} \times \text{power factor}$$
$$= 1500 \times 0.78 = 1170$$
$$(b)\ \mathrm{kvar} = \mathrm{kVA} \times \sin \Theta$$

Now we will use our trigonometric table. The cosine of 39° is 0.7771, which is the value nearest 0.78. (More complete tables would give a more nearly exact value.) Hence the angle between the kVA and the kW is approximately 39°. The sine of 39° is 0.6293, so that

$$\mathrm{kvar} = 1500 \times 0.6293$$
$$= 944.95 \quad \text{or, rounded, } 945$$

TABLE A-1-1 Abbreviated Table of Natural Trigonometric Functions

Angle	Sin	Tan	Cot	Cos	Deg
0	0.0000	0.0000	∞	1.0000	90
1	0.0175	0.0175	57.2900	0.9998	89
2	0.0349	0.0349	28.6363	0.9994	88
3	0.0523	0.0524	19.0811	0.9986	87
4	0.0698	0.0699	14.3007	0.9976	86
5	0.0872	0.0875	11.4300	0.9962	85
6	0.1045	0.1051	9.5144	0.9945	84
7	0.1219	0.1228	8.1443	0.9925	83
8	0.1392	0.1405	7.1154	0.9903	82
9	0.1564	0.1584	6.3138	0.9877	81
10	0.1736	0.1763	5.6713	0.9848	80
11	0.1908	0.1944	5.1446	0.9816	79
12	0.2079	0.2126	4.7046	0.9781	78
13	0.2250	0.2309	4.3315	0.9744	77
14	0.2419	0.2493	4.0108	0.9703	76
15	0.2588	0.2679	3.7321	0.9659	75
16	0.2756	0.2867	3.4874	0.9613	74
17	0.2924	0.3057	3.2709	0.9563	73
18	0.3090	0.3249	3.0777	0.9511	72
19	0.3256	0.3443	2.9042	0.9455	71
20	0.3420	0.3640	2.7475	0.9397	70
21	0.3584	0.3839	2.6051	0.9336	69
22	0.3746	0.4040	2.4751	0.9272	68
23	0.3907	0.4245	2.3559	0.9205	67
24	0.4067	0.4452	2.2460	0.9135	66
25	0.4226	0.4663	2.1445	0.9063	65
26	0.4384	0.4877	2.0503	0.8988	64
27	0.4540	0.5095	1.9626	0.8910	63
28	0.4695	0.5317	1.8807	0.8829	62
29	0.4848	0.5543	1.8040	0.8746	61
30	0.5000	0.5774	1.7321	0.8660	60
31	0.5150	0.6009	1.6643	0.8572	59
32	0.5299	0.6249	1.6003	0.8480	58
33	0.5446	0.6494	1.5399	0.8387	57
34	0.5592	0.6745	1.4826	0.8290	56
35	0.5736	0.7002	1.4281	0.8192	55
36	0.5878	0.7265	1.3764	0.8090	54
37	0.6018	0.7536	1.3270	0.7986	53
38	0.6157	0.7813	1.2799	0.7880	52
39	0.6293	0.8098	1.2349	0.7771	51
40	0.6428	0.8391	1.1918	0.7660	50
41	0.6561	0.8693	1.1504	0.7547	49
42	0.6691	0.9004	1.1106	0.7431	48
43	0.6820	0.9325	1.0724	0.7314	47
44	0.6947	0.9657	1.0355	0.7193	46
45	0.7071	1.0000	1.0000	0.7071	45
Deg	Cos	Cot	Tan	Sin	Angle

2. If watt and var metering was provided, the following information might be available. Load in the circuit is 1400 kW and 900 kvar leading. What are the total kVA and the power factor of the circuit? A diagram of the values would be as shown below:

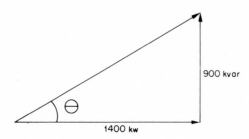

Of course the problem could be solved by squaring the two quantities and taking the square root of the sum, or

$$kVA = \sqrt{(1400)^2 + (900)^2}$$

but it can be done with our trigonometric tables, or by using a calculator which provides trigonometric functions, as follows:

$$\tan \Theta = \frac{900}{1400} = 0.6428$$

From the table, the angle whose tangent is 0.6428 is approximately 33°. The kVA can be found by dividing either the kW by the cosine, or the kvar by the sine of 33°. From the table the cosine of 33° is 0.8387, and the sine is 0.5446. Then

$$kVA = \frac{kW}{\cos \Theta} = \frac{1400}{0.8387} = 1670$$

$$kVA = \frac{kvar}{\sin \Theta} = \frac{900}{0.5446} = 1665$$

With more complete tables showing values of fractions of degrees, the answers found using the two methods of calculation would have been identical. But it should be apparent that the use of trigonometric tables is much simpler than finding the square root of the sum of the squares. However, a calculator makes the solution using the square root of the sum of the squares very simple.

We might look at another slightly more complex problem. If power flows on a bus are as shown in Fig. A-1-5, where the lines are provided with watt and var meters but with no metering on the transformer bank, find the bank load (kW and kVA) and power factor.

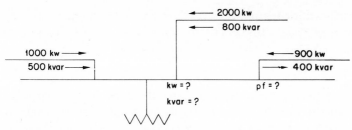

figure A-1-5   Single-line power-flow diagram of three circuit one-bank station.

Since the power (watts) is flowing toward the station in all three circuits, they can be added directly, or

$$\text{Total power} = 1000 + 2000 + 900 = 3900 \text{ kW}$$

The kvar can be added algebraically and would be

$$\text{Total kvar} = 500 + 800 - 400 = 900$$

With a bank load of 3900 kW and 900 kvar determined, it is a simple problem to determine the total kVA and the power factor. The power diagram is as shown below:

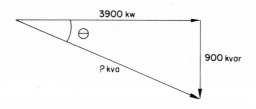

The angle Θ is the angle whose tangent is

$$\frac{900}{3900} = 0.23$$

From the tangent column in the trigonometric table, the angle whose tangent is 0.23 is approximately 13°. The cosine of 13° is 0.9744, so that the power factor is 97.44 percent.

Since the reactive and power are flowing into the bus, the power factor is lagging.

The kVA is the kW load divided by the cosine (PF) of the angle Θ, or

$$\text{kVA} = \frac{3900}{0.9744} = 4005$$

Trigonometric scales are also included on slide rules, so that it is very simple to find appropriate answers to many problems with these devices.

SUMMARY

There is a great deal more to trigonometry than has been covered in this section, but this simple introduction to the subject should make it easy to solve many day-to-day problems in power system operation.

About Vectors

INTRODUCTION

Vectors are used by electrical engineers in solving many problems in ac systems. There is nothing complex about them, and an introduction to vector symbolism should be of value to power system operators in understanding some of the behavior of power systems and circuits.

A vector is merely a means of expressing a quantity and a direction. The quantity can be force, wind, electrical quantities, or almost anything measurable.

Mathematicians distinguish between scalar quantities, which have magnitude only, and vector quantities, which have both magnitude and direction. For example, six apples or a dozen eggs are scalar quantities, but a wind of 10 miles per hour blowing from the north is a vector quantity.

By convention, a vector represented as a horizontal line going to the right is considered positive, and one going horizontal to the left is negative; a vertical line going up is positive, and one going down is negative.

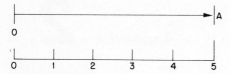

figure A-2-1(*a*) The arrow indicates a vector 5 units long and going to the right or in the positive direction.

figure A-2-1(*b*) The arrow indicates a vector 5 units long but going to the left or in the negative direction.

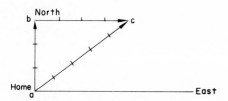

figure A-2-2 Vector representation of paths taken and final location when a man walks 3 mi north and 4 mi east. The final distance from home can be determined as follows:

$$\sqrt{3^2 + 4^2} = \sqrt{9 + 16}$$
$$= \sqrt{25} = 5 \text{ mi}$$

We have determined the distance but still have to determine the direction to fully define the vector giving the final location. From the section on trigonometry we know that the side of the triangle *bc* divided by the side *ab* is the tangent of the angle Θ, or

$$\frac{bc}{ab} = \frac{4}{3} = 1.33$$

From the tangent column of the table of trigonometric functions, the angle whose tangent is 1.33 is approximately 53°. The man's final location is now completely defined as a distance of 5 mi at an angle of 53° east of north.

166

The direction of a vector is indicated by an arrowhead on the end away from the point of origin. Examples are illustrated in Fig. A-2-1. In fact, vectors can go in any direction in space. Three-dimensional representation is a little more complex, but, fortunately, for problems dealing with ac power systems, only two dimensions are necessary. This discussion will therefore be limited to two-dimensional planes.

A simple real-life example might be to determine the actual distance from home (point *a*) that a man would be if he walked 3 miles north (to point *b*) and 4 miles east (to point *c*). This problem is shown vectorially in Fig. A-2-2.

APPLICATION OF VECTORS
TO ELECTRIC POWER SYSTEM PROBLEMS

In electrical problems involving currents and voltages or watts and voltamperes, one of the quantities is usually taken as a reference vector (horizontally to the right), and other quantities are expressed as vectors with an angular displacement to the reference. For currents and voltages the vector system is assumed to be a rotating system with the vectors revolving counterclockwise at synchronous electrical speed. In reality a vector diagram of electrical quantities is a "still" picture of the conditions as they exist at one instant. As we are normally concerned with 60-Hz systems, in 1/60 s a voltage or current vector will make one complete revolution. If, at zero time, a voltage vector is horizontal to the right, in 1/240 s it will be vertical upward (90°), in 1/120 s it will be horizontal in a negative direction, in 1/80 s it will be vertical downward, in 1/60 s it will once again be at the starting position and will repeat this sequence. Vector rotation is illustrated in Fig. A-2-3.

If the vertical projection of a revolving vector is plotted against time of rotation, the resultant plot will be a sine wave, as shown in Fig. A-2-4.

It is from this concept of revolving vectors that the terms "lag" and

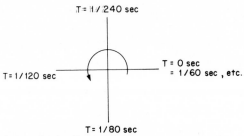

figure A-2-3 Diagram indicating rotation of vectors in a 60-Hz electrical system.

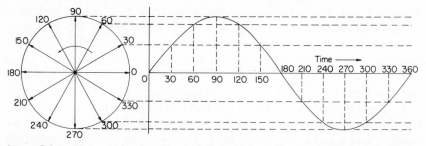

figure A-2-4 Diagram showing the vertical projections of a rotating vector at various time intervals. The curve is the same one that would result if the values of sines of an angle were plotted on a horizontal axis. As a revolving vector will complete one revolution in a given time, time can be expressed as an angle. This is common practice in electrical engineering.

"lead" in power systems are derived. Since currents are frequently displaced in time from voltages because of inductance or capacitance in the circuit, the current vector may not arrive at the time-zero point at the same instant that the voltage arrives at this point. If the current arrives later (inductive case), the current is said to lag the voltage; or if it arrives ahead of the voltage (capacitive case), it is said to lead the voltage. Voltage is ordinarily used as the reference because most systems are constant voltage, although either current or voltage could actually be used as the

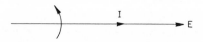

figure A-2-5(a) Current and voltage in phase.

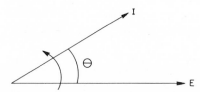

figure A-2-5(b) Current leading voltage.

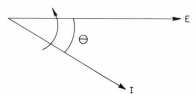

figure A-2-5(c) Current lagging voltage.

reference. When a current arrives at the time-zero point at exactly the same time as the voltage, it is "in phase"; if it is ahead of the voltage, it is "out of phase leading"; and if it arrives later, it is "out of phase lagging." Figure A-2-5 illustrates these conditions.

A practical problem might be to determine the power (watts) and reactive (var) in a single-phase circuit when the current, voltage, and angle between them are known. We know that power is the voltage times the in-phase current, and the reactive is the voltage times the current 90° out of phase. Assume a 120-V circuit, predominantly inductive, with 15 A flowing lagging by 30°. The vector diagram would be as shown in Fig. A-2-6.

We know that the power is the product of the voltage and the in-phase current, and the reactive is the product of the voltage and the out-of-phase component. Using information from the section on trigonometry, it is a simple problem to determine these components. The in-phase component is the total current times the cosine (power factor) of the angle Θ, and the out-of-phase component is the total current times the sine of the angle Θ. From the table of trigonometric functions, the cosine of 30° is 0.8660 and the sine is 0.5000. The power is calculated as follows:

$$\text{Power} = 120 \text{ V} \times 15 \text{ A} \times 0.8660 = 1558.8 \text{ W}$$
$$\text{Reactive} = 120 \text{ V} \times 15 \text{ A} \times 0.5000 = 900.0 \text{ var}$$

Vectors can be added, subtracted, multiplied, and divided as scalar numbers can, but certain rules must be followed. For addition or subtraction, each vector must be broken down into its horizontal and vertical components; these can then be added algebraically and the new resultant vector determined from the sum of the components. A practical example might be several known currents flowing from a bus with the power factors of each known. The problem is to determine the total current on the secondary winding of the transformer supplying this total load. The problem is illustrated in Fig. A-2-7.

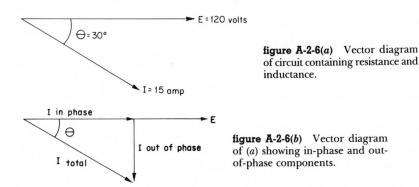

E = 120 volts

Θ = 30°

I = 15 amp

figure A-2-6(a)   Vector diagram of circuit containing resistance and inductance.

I in phase

Θ

E

I out of phase

I total

figure A-2-6(b)   Vector diagram of (a) showing in-phase and out-of-phase components.

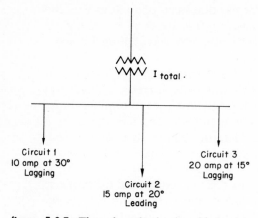

figure A-2-7 Three branch circuits with known currents and angles.

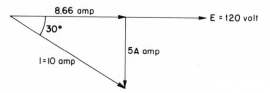

figure A-2-8(a) Vector diagram of circuit 1.

$$I_1 \text{ in phase} = 10 \times \cos 30° = 10 \times 0.8660 = \quad 8.66 \text{ A}$$
$$I_1 \text{ out of phase} = 10 \times \sin 30° = 10 \times 0.5000 \quad = -5.00 \text{ A}$$

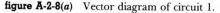

figure A-2-8(b) Vector diagram of circuit 2.

$$I_2 \text{ in phase} = 15 \times \cos 20° = 15 \times 0.9397 = 14.1 \text{ A}$$
$$I_2 \text{ out of phase} = 15 \times \sin 20° = 15 \times 0.3420 = \quad 5.13 \text{ A}$$

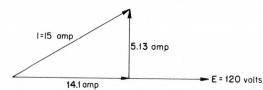

figure A-2-8(c) Vector diagram of circuit 3.

$$I_3 \text{ in phase} = 20 \times \cos 15° = 20 \times 0.9659 = \quad 19.32 \text{ A}$$
$$I_3 \text{ out of phase} = 20 \times \sin 15° = 20 \times 0.2588 = -5.18 \text{ A}$$

The solution to the problem is as follows. The vector diagrams of the circuits are shown in Fig. A-2-8.

Since we have the components, it is now easy to determine the resultant total current and its phase angle.

The algebraic sum of the horizontal components is

$$8.66 + 14.1 + 19.32 = 42.08 \text{ A}$$

The algebraic sum of the vertical components is

$$-5.00 + 5.13 - 5.18 = -5.05 \text{ A}$$

(*Note:* The minus ahead of the current value indicates that the current is lagging the voltage in time.)

We can now completely define the resultant total current whose vector diagram is shown in Fig. A-2-9.

The angle Θ_T is the angle whose tangent is $5.05/42.08 = 0.1190$. From the table of trigonometric functions, the angle whose tangent is 0.1190 is approximately 6°. (This could be interpolated to a more precise value.) The cosine of 6° is 0.9945, so that the total current is $42.08/0.9945 = 42.3$ A lagging by 6°.

The total power taken by the three circuits would be the product of the total in-phase current multiplied by the voltage, or

$$42.08 \times 120 = 5049.6 \text{ W}$$

figure A-2-9   Vector diagram of total current with respect to voltage.

The total reactive would be the product of the total reactive component multiplied by the voltage, or

$$5.05 \times 120 = 606 \text{ var}$$

COMPLEX NUMBERS

Electrical engineers use a very simple notation to describe current, voltage, and impedance vectors, but unfortunately it is called "complex." If a vector *A* of length 1 in the positive direction is multiplied by -1, it will not be changed in length but will be reversed in direction (rotated 180°), as shown in Fig. A-2-10.

If we can find a number which when multiplied by itself is equal to -1, we should be able to rotate the vector by 90°. Because $\sqrt{-1} \times \sqrt{-1} = -1$, it qualifies, but $\sqrt{-1}$ is called an imaginary or complex number. If, however, we consider $\sqrt{-1}$ as a device (operator) for ro-

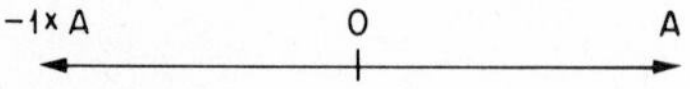

$-1 \times A$ O A

figure A-2-10   Diagram illustrating the reversal of direction of vectors by multiplying by -1.

tating vectors by 90°, its application is very simple. Rather than writing $\sqrt{-1}$ each time, it is customary to use the small letter j (mathematicians use the letter i, but in electrical symbolism this letter is used for current). In the problem whose vector diagrams are shown in Figs. A-2-8 and A-2-9, this form of notation would be applied as follows:

$$I_1 = 8.66 - j5.00$$
$$I_2 = 14.10 + j5.13$$
$$I_3 = \underline{19.32 - j5.18}$$
$$I \text{ total} = 42.08 - j5.05 \qquad \text{the same as previously determined}$$

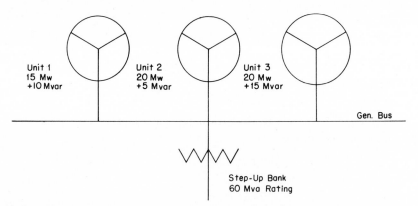

figure A-2-11   Single-line diagram of three-unit generating station with single 60-MVA step-up bank. The problem is to determine the MVA load on the bank.

Unit 1 15 MW + j10 Mvar
Unit 2 20 MW + j 5 Mvar
Unit 3 $\underline{20}$ MW + $\underline{j15}$ Mvar
Total Load 55 MW + j30 Mvar

The rule, then, is to group the horizontal (real) terms and the vertical [reactive (j)] terms and to add the horizontal terms algebraically and then the vertical terms algebraically. The sum is the horizontal and vertical components of the resultant vector. The magnitude and angle of the resultant vector are then easily determined, as demonstrated in the above example.

Problems usually encountered by power system operators are to determine the total MVA load when MW and Mvar quantities are known. The situation illustrated in Fig. A-2-11 might be an example.

The angle whose tangent is $^{30}/_{55} = 0.545$ is 29°. The power factor (cos 29°) is 0.8746. Thus,

$$MVA = \frac{MW}{PF} = \frac{55}{0.8746} = 63 \text{ MVA}$$

or the bank is being operated at 3 MVA above rating.

$$\frac{63}{60} \times 100 = 105\% \text{ of rating}$$

VECTOR MULTIPLICATION AND DIVISION

Although the types of problems encountered by power system operators do not frequently call for it, vectors can easily be multiplied and divided. A vector described by

$$5 + j3$$

can be denoted in what is called polar form by $5\underline{/37°}$. This merely means that the vector is 5 units long at a counterclockwise angle of 37° from the horizontal, as shown in Fig. A-2-12.

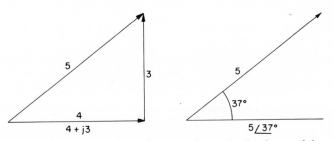

figure A-2-12 Comparison of rectangular and polar forms of describing vectors.

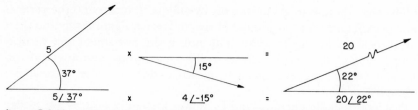

figure A-2-13   Product of two vectors given in polar form is the product of the magnitudes at an angle which is the algebraic sum of the angles of the individual vectors. (*Note:* Vector multiplication and division can also be done in the rectangular form, but it is much more cumbersome.)

No attempt will be made to go through the proof, but if two or more vectors are given in or converted to polar form, their product can be obtained by multiplying the lengths and algebraically adding the angles. Figure A-2-13 illustrates this.

The reverse process can be used to divide vectors; that is, divide one magnitude by the other, and take the algebraic difference of the angles. In the problem just given, if we divide the resultant vector ($20\underline{/22°}$) by the component vector ($4\underline{/-15°}$), we should obtain the original vector ($5\underline{/37°}$).

$$\frac{20\underline{/22°}}{4\underline{/-15°}} = 5\underline{/22° - (-15°)} = 5\underline{/37°}$$

THREE-PHASE REPRESENTATION

In three-phase circuits, the voltage vector of one phase is taken as the reference (horizontal to the right), and other vectors are expressed relative to it, as shown in Fig. A-2-14.

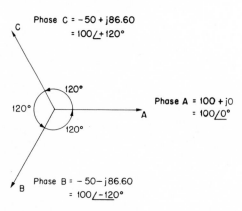

Phase C = −50 + j86.60
    = 100$\underline{/+120°}$

Phase A = 100 + j0
    = 100$\underline{/0°}$

Phase B = −50 − j86.60
    = 100$\underline{/-120°}$

figure A-2-14   Vector representation of balanced three-phase system.

SUMMARY

The above brief discussion does not by any means completely cover the subject of vectors. However, the information provided should prove sufficient to assist power system operators in solving the types of problems most frequently encountered in power system operation.

Revolving Fields

INTRODUCTION

All rotating electric generators and motors make use of the principle that when a conductor is moved in a magnetic field, a voltage (electric potential) will be induced (generator action), or if a conductor carrying electric current is placed in a magnetic field, a force will be produced to move the conductor out of the field (motor action).

In any polyphase generator, when the field winding is moved, the voltages induced in the armature winding will be in those armature coils adjacent to the field poles, as shown in Fig. A-3-1.

VOLTAGE WAVES

The voltages induced in generator windings vary in magnitude with the position of the field poles relative to the armature windings, being a maximum when the field poles are directly adjacent to the armature windings, going to zero when the field poles are 90 electrical degrees

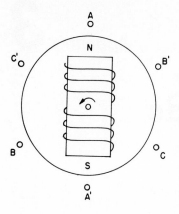

figure A-3-1 Simple representation of a three-phase ac generator. With the field in the position indicated, voltages will be induced in the A-phase winding indicated by A–A'. As the field is revolved, the pole pieces will sweep adjacent to the other phase windings B–B' and C–C', causing voltages to be induced in them in time sequence.

displaced, and being a maximum of the opposite polarity when the opposite field pole sweeps past the coil. The resultant voltages, if plotted against time, will be as shown in Fig. A-3-2.

The shape of the voltage waves shown in Fig. A-3-2 is identical with that which would be produced if trigonometric sines were plotted with

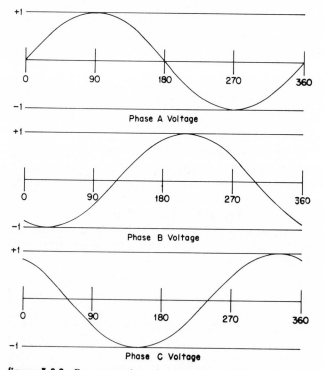

figure A-3-2 Representation of voltages induced in the three-phase windings of a three-phase ac generator versus time.

the sine value as the ordinate (vertical) and the angle (degrees) as the abscissa (horizontal). It is for this reason that the normal ac and voltage waves are referred to as "sine waves."

FLUX WAVES

It is well known that the current that flows in a circuit with a fixed impedance and frequency is directly proportional to the voltage impressed on the circuit. Consequently, if the voltage waves are sinusoidal (sine waves), the currents that flow in the associated circuits will also be sine waves (assuming the load to be a linear impedance), and the amount of current will be equal to the voltage at each instant divided by the circuit impedance.

When current flows in a circuit, a magnetic field is developed, due to the current flow, that is directly proportional to the value of the current, or the magnetic field (flux) is

$$\Phi = K i$$

The Greek letter *phi* (Φ) is conventionally used to denote the magnetic flux. K is a constant which depends on the magnetic properties of the iron in the magnetic circuit, the number of turns in the coils of the windings, and the frequency of the circuit.

The small letter i is used to indicate the value of current in the circuit at any instant.

DEVELOPMENT OF REVOLVING FIELDS

With the above information as a starting point, it is not difficult to show that if polyphase voltages are applied to the armature windings of a motor or generator, a revolving magnetic field of constant magnitude will be produced, and that the field will revolve synchronously (at the same electrical speed) with the system. That is, a machine wound for two electrical poles will produce a field revolving at 3600 rpm, one wound for 4 poles at 1800 rpm, etc., on a 60-Hz system.

To demonstrate the development of the synchronously revolving and constant field, let us assume a circuit such that the maximum value of current will produce a field (flux) of one unit maximum. In this case we can draw sine waves similar to those shown in Fig. A-3-2 as flux waves, Fig. A-3-3.

Because it has been specified that the maximum value of each wave is 1, if the curves showing the flux waves have been carefully drawn to

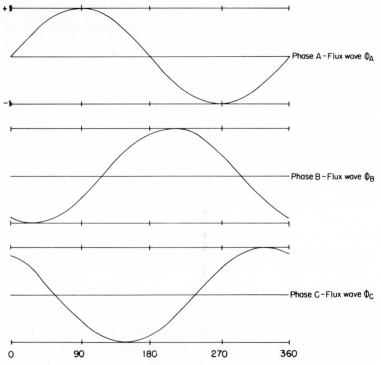

figure A-3-3 Representation of instantaneous values of flux (magnetic fields) developed in armature windings of a three-phase ac machine (generator or motor).

scale, the values at various angles (say 30° intervals) can be read directly from the sine curves for each phase. The sine column of a table of trigonometric functions can also be used if it is remembered that phase B lags phase A by 120° and phase C lags phase A by 240°. The tabulation below gives the values of flux at 30° intervals based on a maximum of 1.

Angle, degrees	Value of A	Value of B	Value of C
0	0	− 0.8660	+ 0.8660
30	+ 0.5000	− 1.000	+ 0.5000
60	+ 0.8660	− 0.8660	0
90	+ 1.000	− 0.5000	− 0.5000
120	+ 0.8660	0	− 0.8660
150	+ 0.5000	+ 0.5000	− 1.000
180	0	+ 0.8660	− 0.8660
210	− 0.5000	+ 1.000	− 0.5000
240	− 0.8660	+ 0.8660	0
270	− 1.000	+ 0.5000	+ 0.5000
300	− 0.8660	0	+ 0.8660
330	− 0.5000	− 0.5000	+ 1.000
360	0	− 0.8660	+ 0.8660

At 360° the cycle will repeat.

By vector addition the magnitude of the resultant total flux vector and its position can readily be determined as shown in Fig. A-3-4.

If the vector additions were continued to 360 electrical degrees, it would be found that the resultant field vector would continue to rotate at synchronous electrical speed and with a constant magnitude.

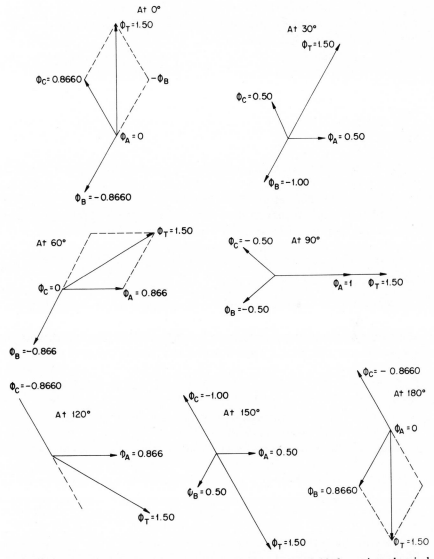

figure A-3-4   Vector diagrams of phase and resultant magnetic fields for various electrical angles at 30° intervals from phase A at 0° to 180°.

REVOLVING FIELDS IN AC MACHINES

The principle of a constant revolving field is very important in understanding polyphase synchronous and induction machines. In paralleling an ac generator to a running system, it is necessary to bring the field poles of the incoming generator to a position such that the revolving field of the incoming machine will produce voltages equal to and in time phase with the running system. This requires that the incoming machine be running at the same electrical speed as the running system. It is only when these conditions exist that the generator switch can be closed. This process is known as synchronizing.

When a polyphase induction motor is connected to a polyphase power source, a revolving field is produced by the currents in the armature windings. This revolving field sweeps past the conductors in the rotor, inducing voltages which cause currents to flow in the rotor conductors. These currents react to produce a torque, causing the rotor to revolve— always at a speed slightly slower than that of the armature revolving field.

When a polyphase synchronous motor is connected to a power source, induction motor action will cause its rotor to revolve as described above. However, as the rotor approaches normal speed, current is applied to the field coils, and by magnetic attraction the rotor is accelerated until it revolves synchronously with the armature revolving field.

TORQUE OR ROTOR ANGLES

It is the position of the armature revolving field at any instant that is used as a reference in determining the torque angles of synchronous machines. When acting as a generator, the machine rotor is driven by an outside power source (prime mover), and as long as the driving force does not exceed a maximum, the rotor field will be advanced slightly from the armature revolving field. The angular difference is the torque angle, and it increases as the load on the machine is increased. If a maximum is exceeded, the magnetic forces will no longer be able to hold the rotor in synchronism, and the machine will go out of step with the running system. The pullout power is affected by the field current; it is greater with a high field current than with a low field current.

When such a machine is used as a motor, when mechanical load is increased, the machine field will drop back with respect to the armature revolving field, and the torque angle will increase until a maximum known as the pullout torque is reached. Any load above the pullout torque will cause the machine to stall. Pullout torque of a synchronous motor is defined as the maximum sustained torque which the motor will

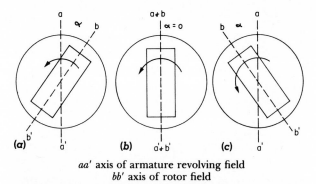

aa' axis of armature revolving field
bb' axis of rotor field

figure A-3-5(a) The rotor field poles lag the position of the armature revolving field by α degrees. Electric energy is absorbed, and mechanical energy is produced (motor action). **(b)** The rotor angle is zero; no mechanical energy is absorbed and no electric energy produced. **(c)** The rotor field is driven ahead of the armature revolving field by the angle α by mechanical-energy input, and electric energy is produced (generator action).

develop at synchronous speed for one minute, with rated voltage applied at rated frequency and with normal excitation. With either motor or generator action, when the torque angles approach the maximum, excessive heating will result. The concept of torque angles for generator and motor action is shown in Fig. A-3-5.

Conversion Factors for Selected Units

1000 milliwatts (mW) = 1 watt (W)
1000 watts = 1 kilowatt (kW)
1000 kW = 1 megawatt (MW)
1 kilowatt = 1.341 horsepower (hp)
1 kilowatt = 3413 British thermal units (Btu)
1 acre-foot = 43,560 cubic feet (ft^3)
1 cubic meter (m^3) = 35.31 cubic feet
1233.6 cubic meters = 1 acre-foot
A flow of 1 cubic feet
per second (cfs) = 1.983 acre-feet per day

Appendix

A

Conversion Factors for
Selected Units

Suggested for Further Study

Kirchmeyer, Leon: *Economic Operation of Interconnected Power Systems,* John Wiley and Sons, Inc., New York, 1959.

Anderson, P. M., and A. A. Fouad: *Power System Control and Stability,* Iowa State University Press, Ames, 1977.

Zaborsky, J., and Joseph Rittenhouse: *Electric Power System Transmission,* Rensselaer Book Store, Troy, N.Y., 1969.

Langsdorf, Alexander S.: *Theory of Alternating Current Machinery,* 2d ed., McGraw-Hill Book Company, New York, 1955.

REA Bulletin 66-10: "Supervisory Control and Energy Management Systems," July 1979.

Westinghouse Electric Corporation: "Applied Protective Relaying," 1979.

Fink, D. G., and H. W. Beaty, eds., *Standard Handbook for Electrical Engineers,* 11th ed., McGraw-Hill Book Company, New York, 1978.

North American Power Systems Interconnection Committee (NAPSIC): "Operating Manual."

Answers to Problems

CHAPTER 1

1. (b)	**2.** (b)	**3.** (b)	**4.** (b)	**5.** (c)
6. (b)	**7.** (c)	**8.** (a)	**9.** (c)	**10.** (d)

CHAPTER 2

1. (b)	**2.** (a)	**3.** (c)	**4.** (c)	**5.** (b)
6. (c)	**7.** (a)	**8.** (a)	**9.** (b)	**10.** (a)
11. (c)	**12.** (b)	**13.** (b)	**14.** (a)	**15.** (c)

CHAPTER 3

1. (b)   **2.** (c)   **3.** (b)
4. Due to the reactances in the circuit
5. (b)   **6.** (a)   **7.** a)53 b)85   **8.** (c)
9. (a) 50Ω   (b) 1500 kvar

CHAPTER 4

1. (c)	**2.** (b)	**3.** (a)	**4.** (b)	**5.** (a)
6. (c)	**7.** (c)	**8.** (b)	**9.** (a)	**10.** (b)	**11.** (d)

CHAPTER 5

1. (c)	**2.** (c)	**3.** (a)	**4.** (b)	**5.** (b)
6. (c)	**7.** (a)	**8.** (c)	**9.** (b)	**10.** (c)

CHAPTER 6

1. (a)	**2.** (c)	**3.** (c)	**4.** (c)	**5.** (b)
6. (a)	**7.** (d)			

CHAPTER 7

1. (a)	**2.** (a)	**3.** (b)	**4.** (c)	**5.** (b)
6. (c)	**7.** (b)	**8.** (c)	**9.** (a)	**10.** (c)
11. (a)	**12.** (c)			

CHAPTER 8

1. (c)	**2.** (c)	**3.** (b)	**4.** (c)	**5.** (d)	**6.** (c)
7. (d)	**8.** (b)	**9.** (c)	**10.** (b)	**11.** (d)	**12.** (a)
13. (b)	**14.** (b)	**15.** (b)	**16.** (a)	**17.** (c)	

CHAPTER 9

1. (c)	**2.** (b)	**3.** (b)	**4.** (a)	**5.** (c)
6. (a)	**7.** (b)	**8.** (c)	**9.** (b)	**10.** (c)
11. (c)	**12.** (b)	**13.** (c)	**14.** (b)	

CHAPTER 10

1. (b)	**2.** (c)	**3.** (b)	**4.** (a)	**5.** (b)
6. (c)	**7.** (a)	**8.** (b)	**9.** (b)	**10.** (b)	**11.** (b)

CHAPTER 11

1. (c)	**2.** (b)	**3.** (c)	**4.** (a)	**5.** (b)
6. (d)	**7.** (c)	**8.** (a)	**9.** (b)	**10.** (a)	**11.** (b)

CHAPTER 12

1. (c)	**2.** (a)	**3.** (c)	**4.** (c)	**5.** (b)
6. (a)	**7.** (b)	**8.** (c)	**9.** (b)	**10.** (b)
11. (a)	**12.** (c)	**13.** (b)	**14.** (c)	**15.** (a)

Index

ABOUT THE AUTHOR

ROBERT H. MILLER has over forty-five years of professional experience in electrical engineering projects. A graduate of Stanford University, he retired after eighteen years as Assistant Manager of Power Control for the Pacific Gas and Electric Company, and spent over four years as a consultant in Latin and South America principally in supervisory control, power line carrier and microwave communications. Currently a private consultant specializing in power system operation, communication, and control, Mr. Miller holds a patent as inventor of an automatic rain gauge. He is the author of various AIEE and IEEE papers and articles in technical journals on communications and power system optimization.